WiMAX

Technologies, Performance Analysis, and QoS

The
WiMAX
Handbook

WiMAX: Technologies, Performance Analysis, and QoS
ISBN 9781420045253

WiMAX: Standards and Security
ISBN 9781420045237

WiMAX: Applications
ISBN 9781420045474

The WiMAX Handbook
Three-Volume Set
ISBN 9781420045350

WiMAX

Technologies, Performance Analysis, and QoS

Edited by
SYED AHSON
MOHAMMAD ILYAS

CRC Press
Taylor & Francis Group
Boca Raton London New York

CRC Press is an imprint of the
Taylor & Francis Group, an informa business

CRC Press
Taylor & Francis Group
6000 Broken Sound Parkway NW, Suite 300
Boca Raton, FL 33487-2742

© 2008 by Taylor & Francis Group, LLC
CRC Press is an imprint of Taylor & Francis Group, an Informa business

No claim to original U.S. Government works
Printed in the United States of America on acid-free paper
10 9 8 7 6 5 4 3 2 1

International Standard Book Number-10: 1-4200-4525-3 (Hardcover)
International Standard Book Number-13: 978-1-4200-4525-3 (Hardcover)

Library of Congress Cataloging-in-Publication Data

Ahson, Syed.
    WiMAX : technologies, performance analysis, and QoS / Syed Ahson and Mohammad Ilyas.
        p. cm.
    Includes bibliographical references and index.
    ISBN 978-1-4200-4525-3 (alk. paper)
    1. Wireless communication systems. 2. Broadband communication systems.
3. IEEE 802.16 (Standard) I. Ilyas, Mohammad, 1953- II. Title.

    TK5103.2.A43215 2008
    621.384--dc22   2007012501

Visit the Taylor & Francis Web site at
http://www.taylorandfrancis.com

and the CRC Press Web site at
http://www.crcpress.com

Contents

Part III QoS

Preface

The demand for broadband services is growing exponentially. Traditional solutions that provide high-speed broadband access use wired access technologies, such as traditional cable, digital subscriber line, Ethernet, and fiber optics. It is extremely difficult and expensive for carriers to build and maintain wired networks, especially in rural and remote areas. Carriers are unwilling to install the necessary equipment in these areas because of little profit and potential. WiMAX will revolutionize broadband communications in the developed world and bridge the digital divide in developing countries. Affordable wireless broadband access for all is very important for a knowledge-based economy and society. WiMAX will provide affordable wireless broadband access for all, improving quality of life thereby and leading to economic empowerment.

The rapid increase of user demands for faster connection to the Internet service has spurred broadband access network technologies advancement over recent years. While backbone networks have matured and are reliable with large bandwidth, the "last mile" remains the bottleneck to enable broadband applications. The driving force behind the development of the WiMAX system has been the desire to satisfy the emerging need for high data rate applications such as voice over IP, video conferencing, interactive gaming, and multimedia streaming. IEEE 802.11-based Wi-Fi networks have been widely deployed in hotspots, offices, campus, and airports to provide ubiquitous wireless coverage. However, this standard is handicapped by its short transmission range, bandwidth, quality of service, and security. WiMAX will resolve the "last-mile" problem in conjunction with IEEE 802.11. WiMAX deployments not only serve residential and enterprise users but may also be deployed as a backhaul for Wi-Fi hotspots or 3G cellular towers.

WiMAX is based on the IEEE 802.16 air interface standard suite, which provides the wireless technology for nomadic and mobile data access. The IEEE 802.16-2004 standard is designed for stationary transmission, and the 802.16e amendment deals with both stationary and mobile transmissions. WiMAX employs orthogonal frequency division multiplexing, and supports adaptive modulation and coding depending on the channel conditions. Wireless systems cover large geographic areas without the need for a costly cable infrastructure to each service access point. WiMAX offers cost-effective and quickly deployable alternative to cabled networks such as fiber-optic links, traditional cable or digital subscriber lines, or T1 networks.

WiMAX offers fast deployment and a cost-effective solution to the last-mile wireless connection problem in metropolitan areas and underserved rural areas. For mass adoption and large-scale deployment of a broadband wireless access system, it must support quality of service (QoS) for real-time and high-bandwidth applications. The IEEE 802.16 standard is a QoS-rich platform. Different access methods are supported for different classes of traffic. Best Effort traffic is one of the most important of these classes as it represents the majority of the overall data traffic. WiMAX employs a reservation-based MAC technology. Reservation-based MAC protocols have been a primary access method for broadband access technologies. For example, the General Packet Radio System, Digital Subscriber Line DSL, and Hybrid Fiber Coaxial HFC cable technologies employ reservation-based multiple access systems.

The WiMAX handbook provides technical information about all aspects of WiMAX. The areas covered in the handbook range from basic concepts to research-grade material including future directions. The WiMAX handbook captures the current state of wireless local area networks, and serves as a source of comprehensive reference material on this subject. The WiMAX handbook consists of three volumes: *WiMAX: Applications*; *WiMAX: Standards and Security*; and *WiMAX: Technologies, Performance Analysis, and QoS*. It has a total of 32 chapters authored by experts from around the world. *WiMAX: Technologies, Performance Analysis, and QoS* includes 10 chapters authored by 28 experts.

Chapter 1 (Design of Baseband Processors for WiMAX Systems) gives an introduction to programmable baseband processors suited for WiMAX systems. Related processing challenges that influence the design of such processors are also highlighted. A mapping of the WiMAX mode IEEE 802.16d onto a programmable processor is used as an example to illustrate the computational requirements on a WiMAX system.

Chapter 2 (Fractal-Based Methodologies for WiMAX Antenna Synthesis) describes an innovative methodology based on perturbed fractal structures for the design of multiband WiMAX antennas. Because of several electrical and geometrical constraints fixed by the project specifications, the synthesis process has been faced with a multiphase Particle Swarm Optimizer-based optimization procedure. The design process as well as the resulting multiband antenna prototype is validated through experimental and numerical tests.

Chapter 3 (Space–Time Coding and Application in WiMAX) addresses and investigates a special class of multiple-input multiple-output, namely, space–time block codes, and its application in WiMAX, the next-generation OFDM-based IEEE 802.16 standard. This chapter describes the principal codes in this class that appear in the IEEE 802.16 standard and its 802.16e-2005 amendment. This chapter demonstrates the value of using multiple antennas and space–time block codes in WiMAX by examining the performance gain of our nonlinear quaternionic code, which utilizes four transmit antennas and achieves full diversity, and comparing it with a single-input single-output implementation.

Chapter 4 (Exploiting Diversity in MIMO-OFDM Systems for Broadband Wireless Communications) reviews space–frequency/space–time–frequency code design criteria and summarizes space–frequency/space–time–frequency coding for multiple-input multiple-output OFDM systems. Different coding approaches for multiple-input multiple-output OFDM systems are explored by taking into account all opportunities for performance improvement in the spatial, the temporal, and the frequency domains in terms of the achievable diversity order.

Chapter 5 (Performance Analysis of IEEE 802.16 Fixed Broadband Wireless Access Systems) proposes a combined call admission control and scheduling scheme for the IEEE 802.16 system considering the bandwidth constraint for the GPSS mode of operation. Analysis of the IEEE 802.16 MAC protocol is done by varying the bandwidth of all the types of services.

Chapter 6 (System Performance Analysis for the Mesh Mode of IEEE 802.16) focuses on the performance analysis of the IEEE 802.16 mesh mode, especially when the transmission is distributed coordinated. Methods for estimating the distributions of the node transmission interval and connection setup delay are developed. The 802.16 mesh medium access control module is implemented in simulator ns-2.

Chapter 7 (Performance Analysis and Simulation Results under Mobile Environments) analyzes the performance of both single-input single-output and space–time block-coded OFDM systems in mobile environments. In the single-input single-output OFDM case, the average bit-error probability is derived. For space–time block-coded OFDM, it is shown that if the channel is time varying, interantenna interference will also be present in addition to intercarrier interference, and the conventional Alamouti detection scheme will fail to detect the transmitted OFDM signal.

Chapter 8 (IEEE 802.16 Multiple Access Control: Resources Allocation for Reservation-Based Traffic) establishes a framework of an ideal reservation period controller, administered at the base station. A two-stage Markov process is used to formulate a Markov Decision Process model, which resembles the dynamics of the reservation-based MAC protocol. A method is established to dynamically calculate the optimized size of the reservation period at the beginning of each frame given the traffic information and state of the system.

Chapter 9 (Scheduling Algorithms for OFDMA-Based WiMAX Systems with QoS Constraints) presents an overview of the ecosystem in which scheduling for IEEE 802.16e systems must be performed. Scheduling problems that can arise due to the nature of subchannelization techniques presented in the 802.16e draft standard are examined and solutions proposed for specific resource allocation problems by abstracting the system as being multitoned and frame based.

Chapter 10 (Resource Allocation and Admission Control Using Fuzzy Logic for OFDMA-Based IEEE 802.16 Broadband Wireless Networks) presents a fuzzy logic controller for admission control in orthogonal frequency division multiple access-based broadband wireless networks. The proposed admission control mechanism considers various traffic source parameters and packet-level quality of service requirement to decide whether an incoming connection can be accepted or not. A queuing model to investigate the impacts of physical layer parameters on the radio link layer queuing performances is formulated.

The targeted audience for the handbook includes professionals who are designers and planners for WiMAX networks, researchers (faculty members and graduate students), and those who would like to learn about this field.

The handbook has the following specific salient features:

- To serve as a single comprehensive source of information and as reference material on WiMAX networks
- To deal with an important and timely topic of emerging communication technology of today, tomorrow, and beyond
- To present accurate, up-to-date information on a broad range of topics related to WiMAX networks
- To present the material authored by the experts in the field
- To present the information in an organized and well-structured manner

Although the handbook is not precisely a textbook, it can certainly be used as a textbook for graduate and research-oriented courses that deal with WiMAX. Any comments from the readers will be highly appreciated.

Many people have contributed to this handbook in their unique ways. The first and the foremost group that deserves immense gratitude is the group of highly talented and skilled researchers who have contributed 32 chapters to this handbook. All of them have been extremely cooperative and professional. It has also been a pleasure to work with Nora Konopka, Helena Redshaw, Jessica Vakili, and Joette Lynch of Taylor & Francis and we are extremely gratified for their support and professionalism. Our families have extended their unconditional love and strong support throughout this project and they all deserve very special thanks.

Syed Ahson
Plantation, FL, USA

Mohammad Ilyas
Boca Raton, FL, USA

Editors

Syed Ahson is a senior staff software engineer with Motorola Inc. He has extensive experience with wireless data protocols (TCP/IP, UDP, HTTP, VoIP, SIP, H.323), wireless data applications (Internet browsing, multimedia messaging, wireless e-mail, firmware over-the-air update), and cellular telephony protocols (GSM, CDMA, 3G, UMTS, HSDPA). He has contributed significantly in leading roles toward the creation of several advanced and exciting cellular phones at Motorola. Prior to joining Motorola, he was a senior software design engineer with NetSpeak Corporation (now part of Net2Phone), a pioneer in VoIP telephony software.

Syed is a coeditor of the *Handbook of Wireless Local Area Networks: Applications, Technology, Security, and Standards* (CRC Press, 2005). Syed has authored "Smartphones" (International Engineering Consortium, April 2006), a research report that reflects on smartphone markets and technologies. He has published several research articles in peer-reviewed journals and teaches computer engineering courses as adjunct faculty at Florida Atlantic University, Florida, where he introduced a course on smartphone technology and applications. Syed received his BSc in electrical engineering from India in 1995 and MS in computer engineering in July 1998 at Florida Atlantic University, Florida.

Dr. Mohammad Ilyas received his BSc in electrical engineering from the University of Engineering and Technology, Lahore, Pakistan, in 1976. From March 1977 to September 1978, he worked for the Water and Power Development Authority, Pakistan. In 1978, he was awarded a scholarship for his graduate studies and he completed his MS in electrical and electronic engineering in June 1980 at Shiraz University, Shiraz, Iran. In September 1980, he joined the doctoral program at Queen's University in Kingston, Ontario, Canada. He completed his PhD in 1983. His doctoral research was about switching and flow control techniques in computer communication networks. Since September 1983, he has been with the College of Engineering and Computer Science at Florida Atlantic University, Boca Raton, Florida, where he is currently associate dean for research and industry relations. From 1994 to 2000, he was chair of the Department of Computer Science and Engineering. From July 2004 to September 2005, he served as interim associate vice president for research and graduate studies. During the 1993–1994 academic year, he was on his sabbatical leave with the Department of Computer Engineering, King Saud University, Riyadh, Saudi Arabia.

Dr. Ilyas has conducted successful research in various areas including traffic management and congestion control in broadband/high-speed communication networks, traffic characterization, wireless communication networks, performance modeling, and simulation. He has published one book, eight handbooks, and over 150 research articles. He has supervised 11 PhD dissertations and more than 37 MS theses to completion. He has been a consultant to several national and international organizations. Dr. Ilyas is an active participant in several IEEE technical committees and activities.

Dr. Ilyas is a senior member of IEEE and a member of ASEE.

Contributors

Naofal Al-Dhahir
University of Texas
Dallas, Texas

Mishal Algharabally
University of California
San Diego, California

Renzo Azaro
University of Trento
Trento, Italy

Robert Calderbank
Princeton University
Princeton, New Jersey

Min Cao
University of Illinois
Urbana-Champaign, Illinois

Jimmy Chui
Princeton University
Princeton, New Jersey

Pankaj Das
University of California
San Diego, California

Sushanta Das
Philips Research N.A.
Briarcliff Manor, New York

Suhas Diggavi
Ecole Polytechnique Fédérale de
  Lausanne
Lausanne, Switzerland

Ahmed Doha
York University
Toronto, Ontario, Canada

Massimo Donelli
University of Trento
Trento, Italy

Hossam Hassanein
Queen's University
Kingston, Ontario, Canada

Ekram Hossain
University of Manitoba
Winnipeg, Manitoba, Canada

McNeil Ivan
Coimbatore Institute of Technology
Coimbatore, India

Raj Iyengar
Rensselaer Polytechnic Institute
Troy, New York

R. Jayaparvathy
Coimbatore Institute of Technology
Coimbatore, India

Koushik Kar
Rensselaer Polytechnic Institute
Troy, New York

Dake Liu
Linköping University and
  Coresonic AB
Linköping, Sweden

K. J. Ray Liu
University of Maryland
College Park, Maryland

Xiang Luo
Rensselaer Polytechnic Institute
Troy, New York

Andrea Massa
University of Trento
Trento, Italy

Anders Nilsson
Linköping University and
  Coresonic AB
Linköping, Sweden

Dusit Niyato
TRLabs
Winnipeg, Manitoba, Canada

Zoltan Safar
Nokia
Copenhagen, Denmark

Biplab Sikdar
Rensselaer Polytechnic Institute
Troy, New York

Weifeng Su
State University of New York
Buffalo, New York

Edoardo Zeni
University of Trento
Trento, Italy

Qian Zhang
University of Science and
  Technology
Hong Kong

Part I

Technologies

1

Design of Baseband Processors for WiMAX Systems

Anders Nilsson and Dake Liu

CONTENTS

1.1   Introduction

A typical wireless communication system contains several signal processing steps. In addition to the radio front-end, radio systems commonly incorporate several digital components such as a digital baseband processor, a media access controller, and an application processor. An overview of such a system is illustrated in Figure 1.1.

Most wireless systems contain two main computational paths, the transmit path and the receive path. In the transmit path, the baseband processor receives data from the media access control (MAC) processor and performs

- Channel coding
- Modulation
- Symbol shaping

before the data is sent to the radio front-end via a digital to analog converter (DAC). In the receive path, the RF signal is first down-converted to an analog baseband signal. The signal is then conditioned and filtered in the analog baseband circuitry. After this the signal is digitized by an analog to digital converter (ADC) and sent to the digital baseband processor that performs

- Filtering, synchronization, and gain control
- Demodulation, channel estimation, and compensation
- Forward error correction (FEC)

before the data is transferred to the MAC protocol layer.

The aim of this chapter is to give an introduction to the programmable baseband processors suited for WiMAX systems and other multimode wireless systems. Related processing challenges that influence the design of such processors are also highlighted. A mapping of the WiMAX mode IEEE

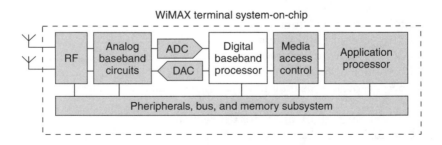

FIGURE 1.1
Radio system overview.

802.16d onto a programmable processor is used as an example to illustrate the computational requirements on a WiMAX system.

1.2 Baseband Processing Challenges

Baseband processing presents a number of challenges that need to be addressed in the design of a wireless system. There are five general challenges that are common to most wireless systems:

- Multipath propagation and fading (intersymbol interference [ISI])
- High mobility
- Frequency and timing offset
- Noise and burst interference
- Large dynamic range

Compensation and handling of these five challenges impose a heavy computational load for the digital baseband processor. Besides the above-mentioned challenges, baseband processing, in general, also faces the challenge of limited computing time and hard real-time requirements.

1.2.1 Multipath Propagation

The data transported between a transmitter and a receiver in a wireless system are affected by the surrounding environment. This gives rise to one of the greatest challenges in wide-band radio links, the problem of multipath propagation and ISI. Multipath propagation occurs when there are more than one propagation paths from the transmitter to the receiver. ISI occurs because all the delayed multipath signal components are added in the receiver. Some frequencies will add constructively and some destructively, since the phases of the received signals depend on the environment. This destroys the original signal. Multipath propagation is illustrated in Figure 1.2.

1.2.2 Timing and Frequency Offset

A slight discrepancy can occur between the transmitter and the receiver carrier frequency and the sample rate as the transmitter and the receiver in a wireless system use different reference oscillators. If uncorrected, this difference limits the useful data rate of a system. In addition, Doppler-spread, which is a frequency-dependent frequency offset caused by mobility, further increases the frequency offset.

1.2.3 Mobility

Mobility in a wireless transmission causes several effects, both Doppler-spread and rapid changes of the channel. The most demanding effect to

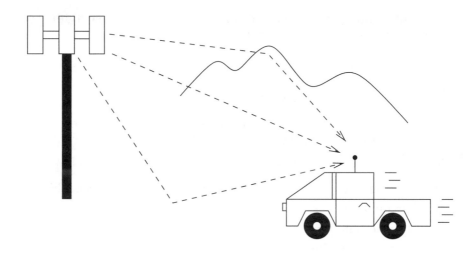

FIGURE 1.2
Multipath propagation.

manage is the rate at which the channel changes. If the mobility is low, e.g., when the channel can be assumed to be stationary for the duration of a complete symbol or data packet, the channel can be estimated by means of a preamble. However, if mobility is so high that the channel changes are significant during a symbol period, then this phenomenon is called fast fading. Fast fading requires the processor to track and recalculate the channel estimate during reception of the user payload data. Hence, it is not enough to rely on an initial channel estimation performed at the beginning of a packet or a frame.

1.2.4   Noise and Burst Interference

Noise and burst interference will degrade the signal arriving to the receiver in a wireless system. Both man-made noise and a natural phenomenon, such as lightning, will cause signal degradation and possible bit-errors. To increase the reliability of a wireless link, FEC techniques are employed. In addition, interleaving is often used to rearrange neighboring data-bits to even out bit-errors caused by burst interference or frequency selective fading. Popular FEC algorithms and codes are the Viterbi algorithm used for convolutional codes, Turbo codes, or Reed–Solomon codes.

1.2.4.1   Dynamic Range

Another problem faced in wireless systems is the large dynamic range of received signals. Both fading and other equipment in the surroundings will increase the dynamic range of the signals arriving at the radio front-end.

A dynamic range requirement of 60–100 dB is not uncommon. Since it is not practical to design systems with such large dynamic range, automatic gain control (AGC) circuits are used. This implies that the processor measures the received signal energy and adjusts the gain of the analog front-end components to normalize the energy received in the ADC. Since signals falling outside the useful range of the ADC cannot be used by the baseband processor, it is essential for the processor to continuously monitor the signal level and adjust the gain accordingly. Power consumption and system cost can be decreased by reducing the dynamic range of the ADC and DAC as well as the internal dynamic range of the number representation in the digital signal processor (DSP). By using smart algorithms for gain control, range margins in the processing chain can be decreased.

1.2.4.2 Processing Latency

Since baseband processing is a strict hard real-time procedure, all processing tasks must be completed on time. This imposes a heavy peak workload for the processor during computationally demanding tasks, such as channel decoding, channel estimation, and gain control calculations. In a packet-based system, the channel estimation, frequency error correction, and gain control functions must be performed before any data can be received. This may result in over-dimensioned hardware, since the hardware must be able to handle the peak workload, even though it may only occur less than one percent of the time. In such cases, programmable DSPs have an advantage over fixed-function hardware since the programmable DSP can reschedule its computing resources to make use of the available computing capacity all the time.

1.3 Programmable Baseband Processors

Since baseband processing is computationally very heavy, baseband processing solutions have traditionally been implemented as fixed-function hardware.

There are two major drawbacks of using nonprogrammable devices. The first is their low flexibility and short product lifetime. A fixed-function product must be redesigned whenever there is a change in the product specification whereas a programmable solution only needs a software update as long as there are enough computing capacity. As most wireless standards tend to evolve over time, programmability is crucial for equipment manufacturers to keep up the pace with standard makers. The WiMAX specification of which parts of the IEEE 802.16 standard to use will most likely change as new features are added to the IEEE 802.16 standard.

The second drawback of fixed-function devices is the excessive need for hardware resources. Designers seldom use hardware multiplexing techniques for digital baseband processing modules because of the added complexity and long verification time. If the module is not programmable, it cannot dynamically allocate computing resources to the respective algorithm, which implies that each function must be mapped to its own specific hardware.

1.3.1 Multimode Systems

As multimode radio terminals become popular and as certain wireless standards such as IEEE 802.16 use several physical layers within the standard, more attention must be paid to the design of the baseband processing hardware.

A WiMAX home gateway will most certainly support the orthogonal frequency division multiplexing (OFDM) and the orthogonal frequency division multiple access (OFDMA) physical layer toward the WiMAX network while serving the home with Wireless LAN, UWB, and digital cordless telephony. The classical way to design multimode systems is to integrate many separate baseband processing modules, each module covering one standard, to support multiple modes. One large drawback of using multiple nonprogrammable hardware modules is the large silicon area used and the lack of hardware reuse.

The trend is to utilize programmable baseband processors instead of the fixed-function hardware. Then, several standards can be implemented with the same hardware, and the function can be changed by just running a different program [1–4].

In the following sections, we will present baseband-specific features of application-specific instruction set processors (ASIPs) and use the LeoCore DSP family from Coresonic [3] as an example.

ASIPs allow the processor architecture to be optimized toward a quite general application area such as baseband processing. By restricting the processor architecture, several application-specific optimizations can be made. By that definition, a baseband DSP capable of processing millions of symbols per second might not be able to decode an MPEG2 video stream.

Design of application-specific processors is all about selecting the right amount of flexibility. The processor must be flexible enough to any wireless standard including possible updates, but not more flexible than so. However, once a system uses programmable baseband processors, several advantages emerge as described in the following sections.

1.3.2 Dynamic MIPS Allocation

By being able to dynamically redistributing available resources among baseband processing tasks, we can focus on either mobility management or high data rate. In Figure 1.3, the MIPS capacity floor of the baseband processor is represented by the broken line. During severe fading conditions, the processor runs advanced channel tracking and compensation algorithms to

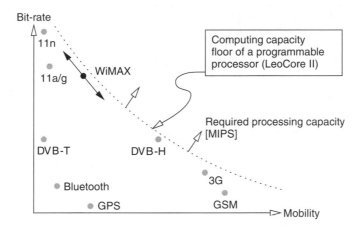

FIGURE 1.3
Dynamic MIPS usage.

provide reliable communication. In good channel conditions, more computing resources can be allocated to symbol processing tasks to increase the throughput of the system. Other common wireless standards are also included in the figure for comparison.

1.3.3 Hardware Multiplexing through Programmability

The IEEE 802.16 standard supports three different kinds of physical layers: OFDM, OFDMA, and single carrier transmission. Normally, WiMAX implies the use of the OFDM or the OFDMA physical layer in nonline of sight conditions and single carrier modulation in microwave back-haul links. This requires the baseband processor to support three different physical layers within the same device. As illustrated in Figure 1.4, all these modulation schemes can be implemented on a single DSP with the functionality shown in the figure.

By carefully selecting the functional blocks, maximum hardware reuse between different standards and modulation schemes can be achieved. As all blocks are programmable, the design is flexible enough to adapt to new standard updates. By partitioning the baseband processing tasks in the five groups shown in Figure 1.4, mapping the tasks to an application-specific processor architecture is simplified. As the processing tasks in most wireless standards can be mapped to these blocks, the processor solution could easily be retargeted to a completely new standard without redesigning the entire processor and software stack. For example, the front-end processing operations are common to all standards, the only differentiating parameters are the filter requirements and sample rate conversion parameters.

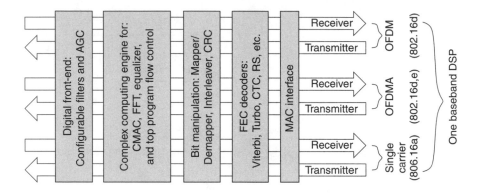

FIGURE 1.4
Hardware multiplexing on LeoCore DSPs.

1.4   IEEE 802.16d Example

To illustrate some of the unique properties of baseband processing and highlight some of the features required in a baseband processor, we analyze the IEEE 802.16d part in the WiMAX specification. As IEEE 802.16d have several modes with different bandwidths and sample rates, we use the most demanding mode in our calculations and figures. However, before the analysis is presented, an introduction to OFDM is given to clarify and motivate some of the operations performed in the processor.

1.4.1   Introduction to OFDM

OFDM is a method that transmits data simultaneously over several subcarrier frequencies. The name comes from the fact that all subcarrier frequencies are mutually orthogonal, thereby signaling on one frequency is not visible on any other subcarrier frequency. This orthogonality is achieved in a nice way in implementation by collecting the symbols to be transmitted on each subcarrier in the frequency domain, and then simultaneously translating all of them into one time-domain symbol using an inverse fast Fourier transform (IFFT).

The advantage of OFDM is that each subcarrier only occupies a narrow frequency band and hence can be considered to be subject to flat fading. Therefore, a complex channel equalizer can be avoided. Instead, the impact of the channel on each subcarrier can be compensated by a simple multiplication to scale and rotate the constellation points to the correct position once the signal has been transferred back to the frequency domain (by way of an FFT) in the receiver.

To further reduce the impact of multipath propagation and ISI, a guard period is often created between OFDM symbols by adding a cyclic prefix (CP) to the beginning of each symbol. This is achieved by simply copying the end of the symbol and adding it in front of the symbol. As long as the channel delay spread is shorter than the CP, the effects of ISI are mitigated.

1.4.2   Processing Job Overview

At this point, it should be quite clear that efficient calculation of the FFT is vital for a baseband processor supporting OFDM. To illustrate the amount of processing needed, Table 1.1 gives an overview of the processing requirements for WiMAX and other well-known radio standards using OFDM for comparison [5–7].

The last line of the table shows the approximate MIPS cost needed only for the FFT itself if the transceiver is implemented in a general DSP processor (with FFT-addressing support). However, FFT is not the only demanding task in an OFDM transceiver. Other heavy jobs adding to the baseband processor requirements are synchronization, channel estimation, and channel decoding.

Figure 1.5 shows typical OFDM processing flows for packet detection/synchronization/channel estimation, for payload reception, and for transmission. Essentially, all processing between mapping/demapping and ADC/DAC manipulates I/Q pairs represented as complex values. The remaining part of the baseband processing consists mainly of channel coding/decoding that typically consists of bit-manipulation operations. Channel coding will be discussed later in this chapter in conjunction with hardware acceleration.

In Table 1.2, benchmarking of key functions in a WiMAX receiver is presented. Along each key function its MIPS cost is presented. The MIPS cost

TABLE 1.1

FFT Computation Complexity

Standard	WiMAX	802.11a/g	DVB-H (4k mode)
Application	Wireless Access	Wireless LAN	Digital TV
Max bit-rate (Mbit/s)	46.6	54	32
Sample rate (MHz)	13.8	20	9.1
FFT size	256	64	4096
Symbol rate (kHz)	43	250	2.2
With radix-2 FFT			
Processing (Mbf/s)	44	48	53
Mem bandwidth (Msample/s)	265	288	319
Memory size (samples)	1536	320	32672
With radix-4 FFT			
Processing (Mbf/s)	11	12	13
Mem bandwidth (Msample/s)	133	144	160
Memory size (samples)	1280	272	26528
Equiv. DSP MIPS	440	480	530

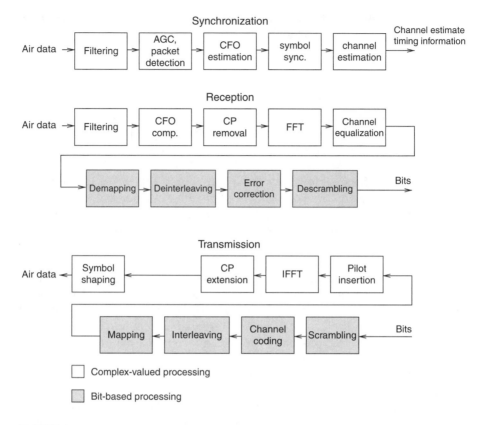

FIGURE 1.5
OFDM processing flow.

TABLE 1.2

Receiver Profiling

Function	Operations	MIPS
Receive/decimation filter	FIR/IIR filter: CMAC	828
Frame detection	Autocorrelation: CMAC	220
Frequency offset estimation	Autocorrelation, complex argument calculation: CMAC, cordic algorithm	70
Frequency offset correction	Rotor: table look-up, CMUL	332
Synchronization	Cross-correlation in time domain: CMAC, absolute maximum *or* in frequency domain: FFT, CMUL, IFFT, absolute maximum	276
Channel estimation	Frequency domain correlation with known pilot symbol: FFT, CMUL	400
Channel equalization	One complex multiplication for each subcarrier: CMUL	120
Demodulation	FFT	440

corresponds to the load of a standard DSP processor with addressing support for FFT-addressing performing the task. After each operation, the main kernel functions are listed.

1.5 Multistandard Processor Design

In this section, a processor architecture suitable for both OFDM, OFDMA and single carrier-based standards, is presented. To summarize the requirements gathered from the previous example, the following points must be considered. The processor must have

1. Efficient instruction set suited for baseband processing. Use of both the natively complex computing and the integer computing.
2. Efficient hardware reuse through instruction level acceleration.
3. Wide execution units to increase processing parallelism.
4. High-memory bandwidth to support parallel execution.
5. Low overhead in processing.
6. Balance between configurable accelerators and execution units.

1.5.1 Complex Computing

A very large part of the processing, including FFTs, frequency/timing offset estimation, synchronization, and channel estimation, all employ well-known convolution-based functions common in DSP processing. Such operations can typically be carried out efficiently by DSP processors, thanks to multiply-accumulate units and optimized memory and bus architectures and addressing modes. However, in baseband processing, essentially all these operations are complex valued. Therefore, it is essential that complex-valued operations can also be carried out efficiently. To reach the best efficency, complex computing should be supported throughout the architecture: by data paths and instruction set as well as by the memory architecture and data types.

1.5.2 Vector Computing

As detailed, application benchmarking shows that most operations in a baseband processor are performed on vectors of complex data, this should be reflected in the instruction set architecture (ISA) of the processor. From the IEEE 802.16d example presented previously, we can confirm that most operations, such as filtering, FFT, and correlation, are performed on large vectors of data. Benchmarking also shows that there are no backward data dependencies among vectors. This enables the possibility to use task pipelines, i.e., pipelining of the entire vector operations. Task level pipelining increases the processing parallelism by running several independent jobs simultaneously

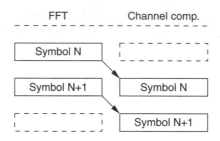

FIGURE 1.6
Task level pipelining.

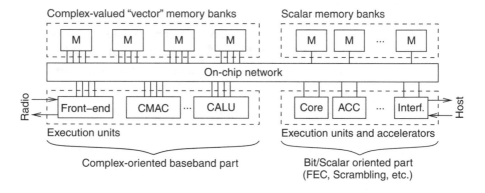

FIGURE 1.7
LeoCore basic architecture.

and passing data between the jobs at specific times. An example of a task level pipeline is shown in Figure 1.6.

The vector property can also be used to improve the addressing efficiency of the processor as most algorithms read and write consecutive data items.

1.5.3 LeoCore Processor Overview

The LeoCore DSP family from Coresonic is used as an example throughout this section. Some of the features and the architecture of the LeoCore DSPs are presented.

The LeoCore DSP consists of two main parts, one natively complex part that mainly operates on vectors of complex numbers and another natively integer part that operates on integers and single bits. The latter part is mainly used for FEC and bit manipulation whereas the former part is used to extract soft data symbols that can be demapped into bits. All units in the processor communicate with each other through an on-chip network, which also provides access to memory banks. An overview of the LeoCore processor architecture is shown in Figure 1.7.

The on-chip network allows any memory to be connected to any execution unit. Execution units span the range from a DSP controller core to multilane complex multiply-accumulate unit (CMAC) and ALU SIMD data paths. Accelerators are also attached to the network.

The architecture relies on the observation that most baseband processing tasks operate on a large set of complex-valued vectors (such as autocorrelation, dot-product, FFT, and convolution). This allows us to optimize execution units to take advantage of this. The LeoCore architecture uses vector instructions, i.e., a single instruction that triggers a complete vector operation such as a complex 128 sample dot-product.

To support this kind of instructions, the execution units must be able to process large data chunks without any intervention from the processor core. This in turn requires the execution unit and memory subsystem to have automatic address generation and efficient load/store subsystems. As a response to this, the base architecture utilizes decentralized memories and memory addressing together with vector execution units.

1.5.4 Single Instruction Issue

A novel feature of the LeoCore DSPs is the way that instructions are issued. As stated earlier, the processor architecture provides vector instructions to improve the processing efficiency. The key idea in the LeoCore architecture is to issue only one instruction each clock cycle while letting several operations execute in parallel as vector instructions may run for several clock cycles on the execution units. This approach results in a degree of parallelism equivalent to a parallel processor without the need for the large control-path overhead. In this way, the vector property of baseband processing could be utilized to reduce the complexity and thus power and area of the processor. For example, the integer data path could execute the operating system's tasks while the CMAC performs one layer of an FFT and the complex ALU (CALU) performs DC-offset cancellation (vector subtraction). This architecture is named single instruction issue multiple tasks (SIMT). The SIMT principle is presented in Figure 1.8.

As shown in Figure 1.8, only one instruction is issued each clock cycle. However, as vector operations will execute for several clock cycles on the execution units, computing parallelism will be maintained with only a single instruction issue each clock cycle. This restriction simplifies the instruction issue logic and enables compact programs since only narrow instructions are used.

1.5.5 Execution Units

To provide an efficient platform for multistandard baseband processing, a baseband processor must provide several high-throughput execution units capable of executing complex tasks in an efficient manner. The LeoCore family of DSPs utilizes all complex-valued execution units that range from CMAC units capable of executing a radix-4 FFT butterfly in one clock cycle to complex ALUs.

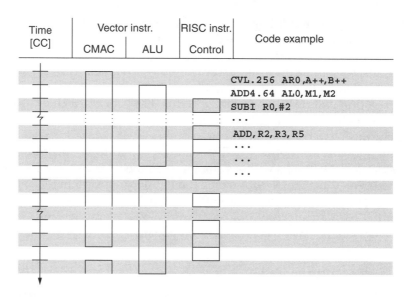

FIGURE 1.8
The principle of single instruction issue multiple tasks (SIMT).

All execution units support vector instructions to support the SIMT technology. Apart from the instruction issue and decoding logic, the execution units are based on the single instruction multiple data (SIMD) principle, i.e., all data paths within the execution unit perform the same operation. By using homogenous SIMD data paths within the execution units and having several different execution units attached to the on-chip network, the processing throughput and diversity are increased while maintaining the high processing efficiency.

1.5.6 Memory Subsystem

The amount of memory needed is often small in baseband processing, but the required memory bandwidth may be very large. As seen in Table 1.1, the FFT calculation alone may need a memory bandwidth of several hundred Msample per second, averaged over the entire symbol time (remember that each sample consists of two values: the real part and the imaginary part). In practice, the peak memory bandwidth required may be up to 1024 bits per clock cycle for a processor running at a few hundred MHz. High-memory bandwidth can be achieved in different ways—using wider memories, more memory banks, or multiport memories—resulting in different trade-offs between flexibility and cost.

Baseband processing is characterized by a predictable flow with a few data dependencies and regular addressing, which means that flexible but expensive multiport memories often can be avoided.

The irregular (bit-reversed) addressing in FFT computations could be considered an exception from this; however, schemes exist that makes it possible to use only the single-port memories and still not cause memory access conflicts even if all inputs/outputs of each butterfly are read/written in parallel.

1.5.7 Hardware Acceleration

To further improve the computing efficiency of the processor, function level accelerators could be used. A function level accelerator is a configurable piece of hardware, which performs a specific task without the support from the processor core.

When deciding which functions to accelerate as function level accelerators, the following must be considered:

MIPS cost: A function with a high MIPS cost may have to be accelerated if the operation cannot be performed by a regular processor.

Reuse: A function that is performed regularly and is used by several radio standards is a good candidate for acceleration.

Circuit area: Acceleration of special functions is only justified if there can be considerable reduction of clock frequency or power compared to the extra area added by the accelerator.

An operation that fulfills one or more of the previous points is a good candidate for hardware acceleration.

1.5.8 FFT Acceleration

Since the FFT is the major corner stone of OFDM processing, it may seem logical to employ a dedicated hardware accelerator block for FFT. Especially since the implementation of such hardware has been widely studied and very efficient solutions exist, e.g., radix-2^2 implementations [8]. However, our experience is that in a programmable solution it is usually more suitable to only accelerate FFT on instruction level by adding butterfly instructions together with bit-reversed/reverse-carry addressing support. There are two main reasons for this:

Flexibility: Many fixed-function FFT implementations tend to lose much of their advantage if multiple FFT sizes must be supported. With butterfly instructions and bit-reversed addressing support, one has full flexibility to efficiently implement any size of FFT. As a bonus, other types of transforms, such as cosine or Walsh transforms, can also be supported.

Hardware reuse: Even the most efficient FFT implementation will contain large hardware components such as (complex valued) multipliers and hence occupy a significant silicon area. If instead one uses dedicated instructions executing in the core data path/MAC unit, these expensive hardware components can be reused by completely different instructions and algorithms. This kind of hardware multi-plexing in fact often means that a programmable solution in many cases can reach a smaller total silicon area than a corresponding fixed-function solution.

1.5.9   Typical Accelerators

1.5.9.1   Front-End Acceleration

In most cases, the received baseband signal will be subject to filtering/decimation in the receiver before it is passed on to the kernel baseband pro-cessing. The required filter can be quite costly in terms of MIPS (again, see Table 1.2). Since this function is needed in almost all radio standards and always runs as soon as the transceiver is in receive mode (receiving data or just waiting for data to appear on the radio channel), the filter is a suitable candidate for acceleration.

Several other functions may also be suitable to include in the same acceler-ator blocks. All these functions are very general and can be reused for many standards:

Resampling: For example, a farrow structure can be used to receive standards with different sample rate using a fixed clock ADC clock *or* to compensate for sample frequency offset between the transmitter and the receiver.

Rotor: A rotor (essentially a numerically controlled oscillator (NCO) and a complex multiplier) can be used to compensate for frequency offset between the transmitter and the receiver. It can also be used for the final down conversion in a low-IF system.

Packet detector: The packet detector recognizes signal patterns that indi-cate the start of a frame. The baseband processor can then be shut down to save power, and be waked up by the packet detector when a valid radio frame arrives.

Shaping filter: During transmission, this filter is used to shape the trans-mitted symbols. This filter is useful in full-duplex systems and can in certain situations be time-shared with the receive filter.

1.5.9.2   Forward Error Correction

FEC functions are also a good candidate for acceleration since they are com-putationally demanding [9] and reused among most WiMAX modes. The WiMAX specification allows a wide variety of FEC algorithms to be used:

- Convolutional codes, Viterbi decoder
- Turbo codes
- Convolutional Turbo codes (CTC)

- Block Turbo codes (BTC)
- Reed–Solomon codes
- Low-density parity check (LDPC) codes

Owing to the variety of allowed FEC algorithms used in WiMAX systems, a flexible accelerator is needed. As most operations can be shared between the Viterbi decoder and Turbo decoder, hardware reuse could be employed in the accelerator.

1.6 Conclusion

This chapter has introduced programmable baseband processors and shown examples of challenges that affect the design of such processors. By moving beyond traditional DSP architectures and introducing new processor architectures, it is possible to implement efficient multistandard baseband processors. The main features of the baseband processor should be

- Inherent support complex-valued computing
- Instruction level acceleration of FFT, convolution, and similar kernel functions
- Optimized memory architecture meeting the high bandwidth and real-time requirements, but typically with a small total amount of memory

In addition, selected tasks should be selected for implementation as accelerators to further improve the computing efficiency. Many of the channel coding tasks as well as some general tasks close to the ADC/DAC interface are often suitable for function level acceleration. Selecting a good trade-off between programmability and function level acceleration ensures versatile yet efficient baseband processors.

References

1. E. Tell, *Design of Programmable Baseband Processors*, PhD thesis, Linköping Studies in Science and Technology, Dissertation No. 969, Linköping, Sweden, Sept. 2005.
2. A. Nilsson, *Design of Multi-Standard Baseband Processors*, Linköping Studies in Science and Technology, Thesis No. 1173, Linköping, Sweden, June 2005.
3. http://www.coresonic.com/, Coresonic AB.
4. http://www.da.isy.liu.se/research, Computer Engineering, Linköping University.
5. IEEE Standard for local and metropolitan area networks, Part 16: Air Interface for Fixed Broadband Wireless Access Systems, WirelessMAN-OFDM.

6. IEEE 802.11a, *Wireless LAN Medium Access Control (MAC) and Physical Layer (PHY) Specifications High-Speed Physical Layer in the 5 GHz Band*, 1999.
7. ETSI EN 300 744, Digital video broadcasting, DVB-T/DVB-H.
8. H. Shousheng and M. Torkelsson, *Designing Pipeline FFT Processor for OFDM (de)Modulation*, Signals, Systems, and Electronics, 1998. ISSSE 98. 1998 URSI International Symposium on 29 Sept.–2 Oct. 1998, pp. 257–262.
9. A. Nilsson and E. Tell, An accelerator structure for programmable multi-standard baseband processors, *Proc. of WNET2004*, Banff, AB, Canada, July 2004.

2

Fractal-Based Methodologies for WiMAX Antenna Synthesis

Renzo Azaro, Edoardo Zeni, Massimo Donelli, and Andrea Massa

CONTENTS

2.1 Introduction

The design of antennas for mobile devices based on wireless standards is a challenging task for the designer and it must be properly dealt with suitable tools and methodologies. In general, the requirements arise from the need for multiband functionalities operations in the limited space of a mobile device. Thus, in addition to severe geometrical constraints, several electrical requirements in terms of impedance matching, antenna gain values, and radiation patterns in multiple frequency bands must be satisfied.

After the application in fixed wireless connections via outdoor radiators and in indoor installation with 802.11-like access point antennas, the wireless technology based on WiMAX standard is expected to be integrated in portable computers and in mobile handset, as well. As a consequence, both geometrical and electrical constraints hold true also for mobile products employing WiMAX standard that support operation in multiple frequencies.

Because of the severe geometrical constraints arising in mobile applications, the antenna designers usually aim at using a single and compact/miniaturized radiating device. However, the allocation of multiple working frequency bands in a single radiator generally turns out to be a complex task with constrained choices.

Generally, the spectral distribution of multiple resonant frequencies in classical antenna geometries [1–3] is strictly related to the geometrical parameters expressed as a function of the wavelength. As an example, elementary wire antennas (i.e., dipoles and monopoles) can work in several frequency bands related to the natural resonances of the structure, but their spectral distribution is regulated by harmonic relationships. Moreover, the electrical properties (voltage standing wave ratio and gain values) vary with the corresponding resonant frequency. The use of a frequency-independent antenna [1–3] may be a solution to comply with multiband requirements. However, the design of a radiator operating only in the requested frequency bands still remains the optimal solution to minimize the effect of out-of-band interfering signals.

In this framework and in recent years, many efforts have been devoted to the design of multiband or wideband miniaturized antennas embedded into the physical structure of a portable device. Monopole-like internal antennas integrated into the radio frequency subsystem and working in the WLAN and WiMAX bands (centered at 2.5 GHz [4] and at 5 GHz [5]) have been proposed. However, dielectric antennas [6] and Yagi arrays [7] have also been used for these applications. Furthermore, a chip antenna [8] for mobile devices that is able to work in the 2.4–2.5 GHz bands in direct contact with electronic components and metallic parts has been studied.

To allow WiMAX services from 2.3 up to 5.9 GHz, a wideband approach has also been considered thus defining a broadband coaxial structure [9], a stubby monopole [10], and a folded planar monopole [11]. As far as the multiband approach is concerned, a multiband monopole antenna for laptop computer [12], multiband dipole antennas [13,14], and a planar inverted-F antenna for USB dongle [15] have been described and realized.

Despite the successful results shown in Refs. 4–15 and obtained through an efficient use of a set of radiating structures derived from classical antenna geometries, the availability of antennas effective in multiple frequency bands and tunable according to the design constraints (both electrical and geometrical) by only varying their geometrical parameters according to a suitable methodology is still an unrealized desire.

A possible answer to such a request is the exploitation of the radiation properties of fractal geometries, both for multiband operations and antenna miniaturization.

In the following, Section 2.2 presents a survey of fractal antenna properties. Then, the solution methodology based on the use of perturbed fractal-like shapes is described in Section 2.3. A set of selected and representative synthesis results concerned with WiMAX applications are presented in Section 2.4. Finally, some conclusions are drawn (Section 2.5).

2.2 Fractal Antenna Properties

Recently, the electrodynamical properties of fractal geometries [16,17] have been extensively studied by several authors [18] focusing on their multiband behavior and ability to operate as efficient small antennas. As a matter of fact, the use of fractal or prefractal geometries (characterized by a finite number of fractal iterations) for antenna synthesis has been proven to be very effective in achieving miniaturized dimensions and an enhanced bandwidth [18,19], even though a reduction of the radiation efficiency at resonant frequencies [20–22] takes place. Some interesting applications have been presented in literature [23,24] confirming that fractal and prefractal geometries are suitable candidates for the synthesis of multiband antennas. However, as pointed out in Ref. 23 dealing with Koch-like fractal geometries, classical fractal or prefractal structures usually present a harmonic dependence rather than a multiband behavior. For such a reason, likewise to classical antennas, the free allocation in the frequency spectrum of working bands noncorrelated through harmonic relationships turns out to be a difficult task also with standard fractal or prefractal shapes.

To overcome such a limitation, an approach is based on the insertion of reactive loads in the antenna structure for obtaining and controlling multiple resonant frequencies at the cost of an increased complexity of the antenna building process [25,26]. As far as fractal geometries are concerned, a similar solution has been used by exploiting the properties of fractals and prefractals combined with an optimization algorithm to optimize both the antenna geometry, the values of reactive loads, and their positions [27].

Another effective methodology for obtaining a multiband behavior, but avoiding the insertion of lumped loads, is based on the addition of other degrees of freedom in the synthesis process by perturbing the fractal geometry. As far as the effects of geometrical perturbations on electrodynamical properties of fractals antennas are concerned, some authors have shown [28,29] that Koch-like and Sierpinski shapes present electrical characteristics similar (or slightly worse) to those of optimized monopolar antennas with an equivalent length. However, it has been shown [29] that perturbed fractal shapes allow performances better than those of the corresponding classical fractal shapes. As a consequence, the introduction of perturbations or modifications of the geometrical parameters of the fractal antenna turns out to be effective for achieving a fruitful exploitation of the radiation properties of fractal-like geometries.

According to these indications, some designs of multiband and miniaturized antennas have been presented in literature. In Ref. 30, a study on the modifications of the spacing among the working bands of a Sierpinki-like antenna has been reported, and in Refs. 31 and 32 the synthesis of dual band antennas working in nonharmonic frequency bands has been described. Moreover, the miniaturization of a monopole antenna in the UHF band has been presented in Ref. 33.

2.3   Synthesis of Fractal-Like Antennas

An effective approach for exploiting miniaturization and multiband properties of fractal or prefractal geometries is based on the introduction of perturbations in reference fractal shapes to increase the number of degrees of freedom of the antenna structure thus allowing a more effective fitting with the electrical requirements in each frequency band.

Accordingly, to comply with electrical and geometrical constraints by properly defining the geometrical parameters (or equivalent degrees of freedom) of the radiating system, the synthesis procedure can be fruitfully recast as an optimization problem.

Starting from the electrical constraints usually expressed as

$$G_b\{\theta, \varphi, f\} \geq G_b^{\min} \qquad \text{VSWR}_b\{f\} \leq \text{VSWR}_b^{\max} \qquad f_b^{\min} \leq f \leq f_b^{\max} \qquad (2.1)$$

(where $b = 1, \ldots, B$ is the index of bth band; f the working frequency; and G and VSWR are the antenna gain and the voltage standing wave ratio, respectively) and under the assumption that the physical antenna is required to belong to a fixed volume

$$
\begin{aligned}
x^{\min} &\leq x_a \leq x^{\max} \\
y^{\min} &\leq y_a \leq y^{\max} \\
z^{\min} &\leq z_a \leq z^{\max}
\end{aligned}
\qquad (2.2)
$$

where $\{x_a, y_a, z_a\}$ identifies a point on the extent of the antenna, it is convenient to define the "solution space" of the optimization problem as follows:

$$G\{\theta, \varphi, f\} = \Phi_a(\underline{\gamma}_a) \geq G^{\min} \quad \text{VSWR}\{f\} = \Psi_a(\underline{\gamma}_a) \leq \text{VSWR}^{\max} \quad f_b^{\min} \leq f \leq f_b^{\max} \tag{2.3}$$

$\underline{\gamma}_a$ being the unknown array coding the descriptive parameters of the antenna structure.

Then, $\underline{\gamma}_a$ is determined by minimizing a suitable cost function

$$
\begin{aligned}
\Omega(\underline{\gamma}_a) = \sum_{i=0}^{F-1} \sum_{v=0}^{V-1} \sum_{t=0}^{T-1} &\left\{ \max\left[0, \frac{\Phi_a(t\Delta\theta, v\Delta\varphi, i\Delta f) - G_a^{\min}}{G_a^{\min}} \right] \right\} \\
&+ \sum_{i=0}^{F-1} \left\{ \max\left[0, \frac{\text{VSWR}_a^{\max} - \Psi_a(i\Delta f)}{\text{VSWR}_a^{\max}} \right] \right\}
\end{aligned}
\qquad (2.4)
$$

where $\Delta\theta$, $\Delta\varphi$, and Δf are sampling intervals, $\Phi_a\{t\Delta\theta, v\Delta\varphi, i\Delta f\} = \Phi_a(\underline{\gamma}_a)$, and $\Psi_a\{i\Delta f\} = \Psi_a(\underline{\gamma}_a)$.

To efficiently explore the solution space and thus minimizing Equation 2.4, a suitable optimization technique is needed. Toward this end, a customized implementation of the particle swarm optimizer (PSO) [34–36] has been integrated with a prefractal geometry generator and a electromagnetic simulator based on the *method of moment* (MoM) [37].

The PSO is a robust stochastic search procedure, inspired by the social behavior of insects swarms, proposed by Kennedy and Eberhart in 1995 [38]. Thanks to its features in exploring complex search spaces, PSO has been employed with success in the framework of applied and computational electromagnetics [34,35] as well as in the field of antenna synthesis [39–41].

In our case, starting from each of the trial arrays $\gamma_m^{(k)}$ (m is the trial array index, $m = 1, \ldots, M$; k the iteration index, $k = 1, \ldots, K$) defined by the swarm-logic, the prefractal generator defines the corresponding prefractal antenna structure. Then, VSWR and gain values are computed by means of the MoM simulator, which takes also into account the presence of the dielectric slab, whether the antenna is printed on a dielectric substrate or the effect of a reference ground plane when a monopolar antenna is considered. The iterative process continues until $k = K$ or $\Omega_{\text{opt}} \leq \eta$ ($\Omega_{\text{opt}} = \min_{k,m}\{F\lfloor\underline{\gamma}_m^{(k)}\rfloor\}$), K and η are the maximum number of iterations and the convergence threshold, respectively.

2.4 Synthesis and Optimization of Miniaturized and Multiband WiMAX Fractal Antennas

To show the effectiveness of the PSO-based synthesis technique in dealing with WiMAX antennas, three test cases concerned with miniaturized and multiband antennas will be described by showing and comparing selected numerical and experimental results. As far as the considered geometries are concerned, only planar structures printed on a dielectric substrate have been taken into account to obtain cheap structures easily embedded into mobile devices.

2.4.1 Synthesis and Optimization of a 3.5 GHz Miniaturized WiMAX Koch-Like Fractal Antenna

Dealing with a system for portable devices, the radiation characteristics of a monopolar quarter-wave-like pattern have been assumed as reference. Because of the broad frequency band required by 802.16 WiMAX applications and according to the European Standard ETSI EN 302 085 V1.2.2 (2003-08), a voltage standing wave ratio lower than $\text{VSWR}_{\text{max}} = 1.8$ (i.e., a reflected power at the input port lower than 10% of the incident power) in the 3.4–3.6 GHz frequency range is required.

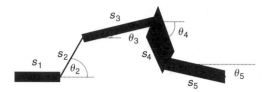

FIGURE 2.1

Descriptive parameters of the Koch-like miniaturized WiMAX antenna.

From a geometrical point of view, a size reduction of more than 20% compared to a standard quarter-wave resonant monopole is needed. Moreover, the antenna belongs to a physical platform of dimensions $L_{max} = 16$ (mm) $\times H_{max} = 10$ (mm).

As far as the building block is concerned, the Koch-like trapezoidal curve proposed in Ref. 27 has been used. Therefore, the antenna structure is uniquely determined by the following descriptive parameters (Figure 2.1): $s_1, s_2, s_4, s_5, \theta_2, \theta_3, \theta_4$ (i.e., the parameters that define the set of affine transformations employed by the iterated function system (IFS) [27] for generating prefractal antenna elements), L (i.e., the projected length of the fractal structure), and w_1, w_2, w_3, w_4, w_5 (i.e., the widths of the fractal segments), which are requested to satisfy the geometrical constraints given by

$$\sum_{i=1}^{I} s_i \cos \theta_i = L \qquad \sum_{i=1}^{I} s_i \sin \theta_i = 0, \qquad I = 5 \qquad (2.5)$$

with $\theta_1 = 0°$.

Furthermore, to avoid the generation of physically unfeasible or very complex solutions, other additional physical constraints have been imposed on the antenna parameters and a penalty has been imposed on those configurations that while not unfeasible would be difficult to realize (e.g., higher fractal orders or large ratio between width and length of the fractal segment).

To satisfy the project guidelines, the unknown array $\underline{\gamma}_a = \{s_1, s_2, s_4, s_5, \theta_2, \theta_3, \theta_4; w_i, i = 1, \ldots, I\}$ is determined by the minimization of $\widetilde{\Omega}(\underline{\gamma}_a)$ (Equation 2.4) in the range 3.4–3.6 GHz.

As far as the PSO parameters are concerned, a population of $M = 15$ trial solutions, a threshold $\eta = 10^{-3}$, and a maximum number of iterations equal to $K = 450$ have been assumed. The remaining parameters of the PSO have been set according to the reference literature [34] as in Ref. 36.

Figure 2.2 shows the evolution of the geometry of the antenna structure during the iterative process. At each iteration, the structure of the best solution (i.e., $\underline{\gamma}_{opt}^{(k)} = \arg(\min_m\{\Omega\lfloor\underline{\gamma}_m^{(k)}\rfloor\})$) is given and the plot of the corresponding VSWR function is illustrated in Figure 2.3a. As it can be observed, starting from a completely mismatched behavior corresponding to the structure shown in Figure 2.2a ($k = 0$), the solution improves until the final shape as shown in Figure 2.2d ($k = k_{conv}$) that fits the requested electrical and geometrical specifications. As a matter of fact, the synthesized structure satisfies the

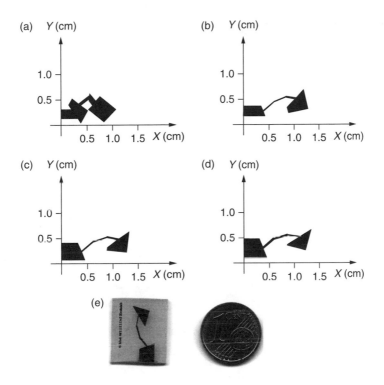

FIGURE 2.2
Koch-like miniaturized WiMAX antenna. Geometry of the antenna at different iteration steps of the optimization procedure: (a) $k=0$, (b) $k=50$, (c) $k=100$, (d) $k=k_{conv}$, and (e) photograph of the prototype.

geometrical requirements since its transversal and longitudinal dimensions are equal to $L_{opt}=13.39$ mm along the x-axis and $H_{opt}=5.42$ mm along the y-axis, respectively. In particular, the projected length L_{opt} turns out to be lower than that of the resonant monopole printed on FR4 substrate, with a reduction equal to 24.77%.

Successively, an antenna prototype has been built on an FR4 substrate by using a photolithographic printing circuit technology and according to the geometric guidelines of the optimized geometry shown in Figure 2.2d. Concerning the VSWR measurements, the antenna prototype (Figure 2.2e) has been equipped with an SMA connector and it has been placed on a reference ground plane with dimensions equal to 90 cm × 140 cm. Moreover, the VSWR has been measured with a scalar network analyzer placing the antenna inside an anechoic chamber.

Computed and measured VSWR values have been compared and the results are shown in Figure 2.3b. As it can be noticed, measured as well as simulated VSWR values satisfy the project specifications in the 3.4–3.6 GHz band. Although a reasonable agreement between the simulation and the

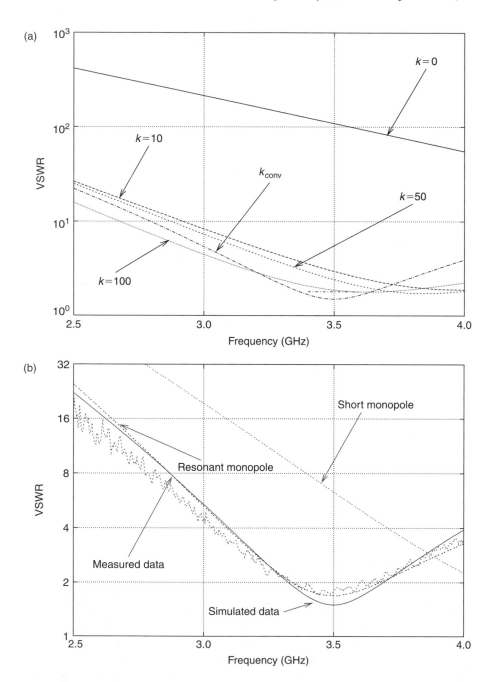

FIGURE 2.3
Koch-like miniaturized WiMAX antenna: (a) simulated VSWR values at the input port at different iteration steps of the optimization procedure and (b) the comparison between measured and simulated values for the synthesized Koch-like antenna and simulated values for reference resonant monopole and short monopole.

experimental results can be observed, some differences occur and the VSWR values measured in the WiMax band turn out to be greater than those simulated. Such a behavior can be attributed to some approximations in the numerical model of the dielectric properties of the FR4 substrate and the ground plane. For comparison purposes, the VSWR values of the prefractal antenna are also compared with those of a resonant quarter-wave monopole ($L_{\lambda/4} = 7.8$ mm long printed on an FR4 substrate—called *resonant monopole*) and a straight monopole with the same length L_{opt} of the WiMAX antenna (called *short monopole*). As expected, the short monopole is not able to fit the VSWR specifications in the working band. However, the simulated values of the resonant monopole seem to indicate some difficulties in satisfying the VSWR constraints.

For completeness, Figure 2.4a plots the horizontal gain function of the WiMAX antenna, while in Figure 2.4b the vertical gain function is shown. As required for a portable wireless device, the radiation properties of the optimized WiMAX fractal antenna are very close to those of a conventional monopole.

2.4.2 Synthesis and Optimization of a Dual-Band WiMAX Koch-Like Fractal Antenna

The second test case deals with the design of a dual-band Koch-like [27] ($f_1 = 2.5$ GHz and $f_2 = 3.5$ GHz) WiMAX antenna. Radiation characteristics that guarantee a hemispherical coverage and a VSWR lower than 1.8 in both the working frequency bands have been assumed. Concerning the geometrical constraints, the antenna belongs to a physical platform of dimensions $L_{max} = 80$ [mm] $\times H_{max} = 40$ [mm]. As far as the general shape of the generating antenna is concerned, a Koch-like geometry has been used according to the notation in Ref. 27, the antenna has been generated from the Koch curve by repeatedly applying the so-called Hutchinson operator until the stage $i = 2$, to achieve two resonant frequencies to be tuned. As shown in Figure 2.5, the antenna structure is uniquely described by a set of segment lengths $L_{i,r,j}$, a set of segment widths $W_{i,r,j}$, and a set of angles $\Theta_{i,r,v}$, where i is the index of the fractal stage, $r = 1, \ldots, R$ the index of the self-similar objects [17,18] that can be considered at the ith stage (i.e., the number of the smaller copies of the generator found at the ith stage), $j = 1, \ldots, J$ the index of the generic segment forming the self-similar object having index r, and $v = 1, \ldots, N$ the index of the bent angles in each self-similar object with index r. Consequently, the unknown descriptive parameters turn out to be $\underline{\gamma}_a = \{L_{i,r,j}; W_{i,r,j}; \Theta_{i,r,v};$ $i = 2; r = 1, \ldots, 4^{i-1}; j = 1, \ldots, 4; v = 1, 2\}$.

To synthesize the antenna geometry, the cost function $\Omega(\underline{\gamma}_a)$ has been minimized by the PSO optimizer with $P = 8$ trial solutions at each iteration. Some representative geometries at various synthesis steps are shown in Figure 2.6a through 2.6d, while in Figure 2.7 the corresponding VSWR functions are reported. In particular, Figures 2.6d and 2.7 ($k = k_{conv}$)

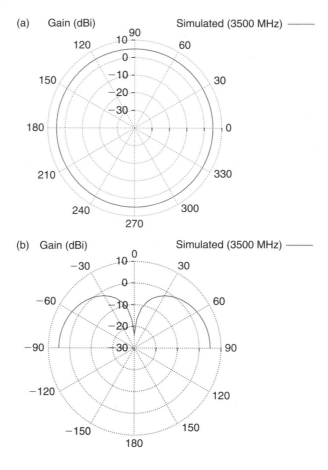

FIGURE 2.4
Koch-like miniaturized WiMAX antenna: simulated gain function at (a) the horizontal plane and (b) at the vertical plane ($\phi = 0°$).

show the geometry and the VSWR of the synthesized dual-band antenna, respectively.

As it can be observed, the synthesized antenna also fits the geometrical constraints since its transversal and longitudinal dimensions are equal to 61 mm along the x-axis and 23 mm along the y-axis.

As far as the experimental measurements are concerned, the antenna prototype (Figure 2.6e) has been equipped with an SMA connector and located over a reference ground plane 90 cm × 140 cm in extension. Computed and measured VSWR values have been compared and the results are shown in Figure 2.7. As expected, there is a good agreement between simulated and measured data. Moreover, the project constraints are satisfied both at $f_1 = 2.5\,\text{GHz}$ and at $f_2 = 3.5\,\text{GHz}$ ($\text{VSWR}_{2.5\,\text{GHz sim}} = 1.13$ versus $\text{VSWR}_{2.5\,\text{GHz max}} = 1.27$, $\text{VSWR}_{3.5\,\text{GHz sim}} = 1.11$ versus $\text{VSWR}_{3.5\,\text{GHz max}} = 1.13$).

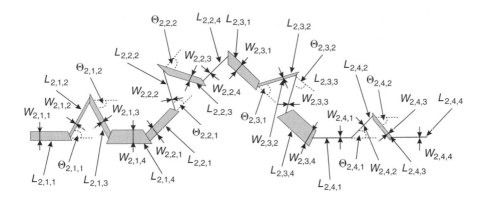

FIGURE 2.5
Descriptive parameters of the Koch-like dual-band WiMAX antenna.

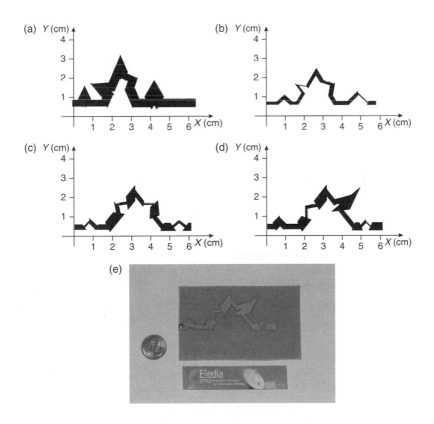

FIGURE 2.6
Koch-like dual-band WiMAX antenna. Geometry of the antenna at different iteration steps of the optimization procedure: (a) $k = 0$, (b) $k = 100$, (c) $k = 200$, (d) $k = k_{conv}$, and (e) photograph of the prototype.

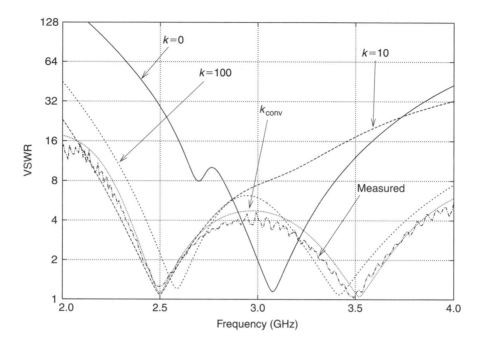

FIGURE 2.7
Koch-like dual-band WiMAX antenna: comparison between simulated VSWR values at the input port at different iteration steps of the optimization procedure and VSWR values measured with the prototype of the optimized antenna.

For completeness, Figure 2.8a and 2.8b shows the simulated gain functions in the horizontal plane ($\theta = 0°$) and in a vertical plane ($\phi = 0°$), respectively, confirming the hemispherical coverage in both the frequency bands.

2.4.3   Synthesis and Optimization of a Dual-Band WiMAX Sierpinski-Like Fractal Antenna

By assuming the same constraints of the previous example (Section 2.3.2) but smaller dimensions ($L_{max} = 40$ [mm] $\times H_{max} = 40$ [mm]) the last example is concerned with a dual-band. The shape of the generating antenna is a Sierpinski-like prefractal geometry [18,30] stopped at the stage $i = 2$ to tune the antenna over two resonant frequencies. As shown in Figure 2.9, the antenna structure is uniquely described by a set of segment lengths $L_{i,r,j}$ and a set of angles $\Theta_{i,r,v}$, where i is the index of the fractal stage, $r = 1, \ldots, R$ denotes the index of the self-similar objects [17,18] that can be considered at the ith stage (i.e., the number of the smaller copies of the generator found at the stage i), $j = 1, \ldots, J$ the index of the generic side in the self-similar object with index r, and $v = 1, \ldots, N$ identifies an angle in each self-similar object with index r.

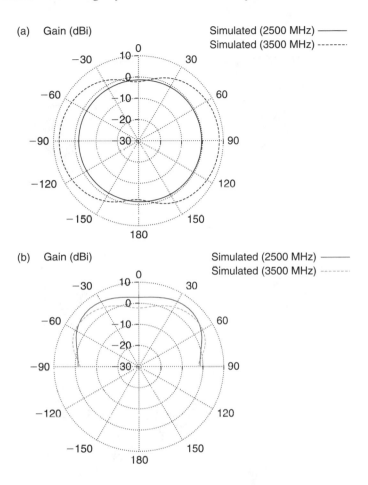

FIGURE 2.8
Koch-like dual-band WiMAX antenna: simulated gain function at (a) the horizontal plane and (b) at the vertical plane ($\phi = 0°$).

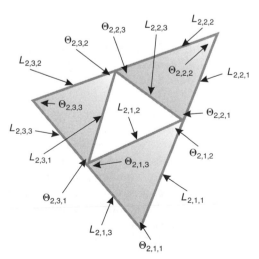

FIGURE 2.9
Descriptive parameters of the Sierpinski-like dual-band WiMAX antenna.

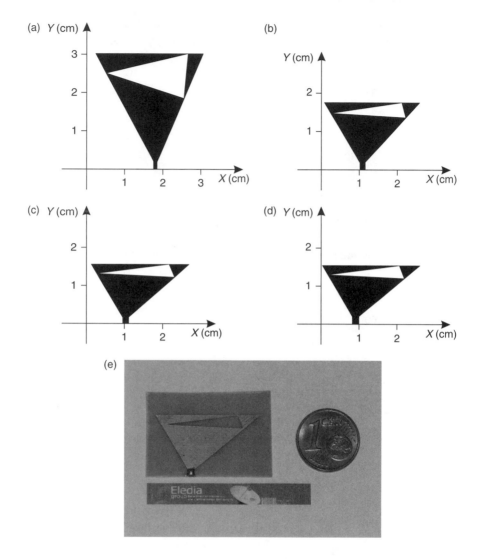

FIGURE 2.10
Sierpinski-like dual-band WiMAX antenna. Geometry of the antenna at different iteration steps of the optimization procedure: (a) $k = 0$, (b) $k = 10$, (c) $k = 100$, (d) $k = k_{conv}$, and (e) photograph of the prototype.

As a result, all the descriptive unknown parameters can be written as $\underline{\gamma}_a = \{\Phi_1; L_{i,r,j}; \Theta_{i,r,v}; i = 2; r = 1, \ldots, 3^{i-1}; j = 1, \ldots, 3; v = 1, 2, 3\}$, where Φ_1 is an orientation angle with reference to the ground plane.

For the synthesis process, the cost function $\Omega(\underline{\gamma}_a)$ has been minimized by considering a smaller population ($P = 5$) of trial solutions because of the complex geometry and the increasing computational costs (e.g., CPU-time for a cost function evaluation). Some representative geometries at various synthesis steps of the iterative process are shown in Figure 2.10a

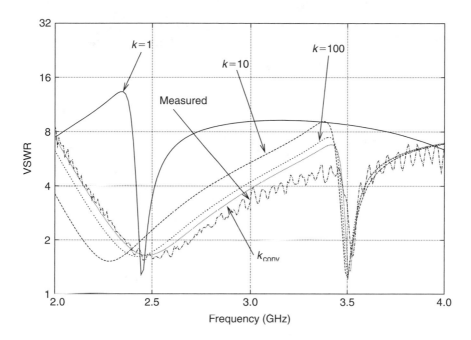

FIGURE 2.11
Sierpinski like dual-band WiMAX antenna: comparison between simulated VSWR values at the input port at different iteration steps of the optimization procedure and VSWR values measured with the prototype of the optimized antenna.

through 2.10d. Moreover, Figure 2.11 gives the plots of the values of the VSWR.

The resulting antenna, whose geometrical and electrical characteristics are given in Figures 2.10d and 2.11 ($k = k_{conv}$), has been built (Figure 2.10e) and the results of the comparative study between computed and measured VSWR values are reported in Figure 2.11 ($VSWR_{2.5\,GHz\,sim} = 1.65$ versus $VSWR_{2.5\,GHz\,max} = 1.56$, $VSWR_{3.5\,GHz\,sim} = 1.34$ versus $VSWR_{3.5\,GHz\,max} = 1.65$).

As far as the gain pattern is concerned, Figure 2.12a and 2.12b shows the simulated gain functions in the horizontal plane ($\theta = 0°$) and in a vertical plane ($\phi = 0°$), respectively.

2.4.4 Computational Issues of the PSO-Based Synthesis Procedure

To give some indications on the computational issue of the PSO-based synthesis procedure, Figure 2.13 displays the plots of the optimal value cost function $F_{(k)}^{opt} = \min_m \{F\lfloor \underline{\gamma}_{-m}^{(k)} \rfloor\}$ versus the iteration number for the synthesis of the various WiMAX antennas described in Sections 2.4.1 through 2.4.3. As it can be noticed, whatever the test case, the convergence value of Ω turns out lower than 10^{-1} pointing out an accurate matching with the synthesis constraints.

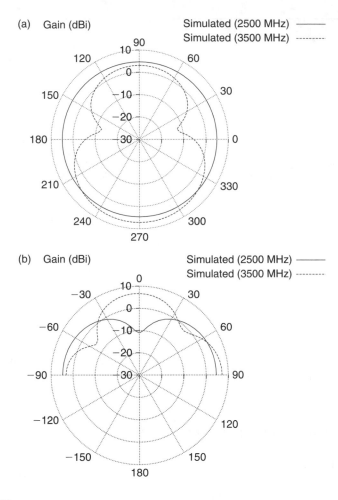

FIGURE 2.12
Sierpinski-like dual-band WiMAX antenna: simulated gain function at (a) the horizontal plane and (b) at the vertical plane ($\phi = 0°$).

However, as expected, antenna structure causes a decrease in the effective-ness of the optimization process also caused by a reduced dimension of the trial population.

2.5 Conclusions

In this work, an innovative methodology based on perturbed fractal struc-tures for the design of multiband WiMAX antennas has been described. Because of several electrical and geometrical constraints fixed by the project specifications, the synthesis process has been faced with a multiphase

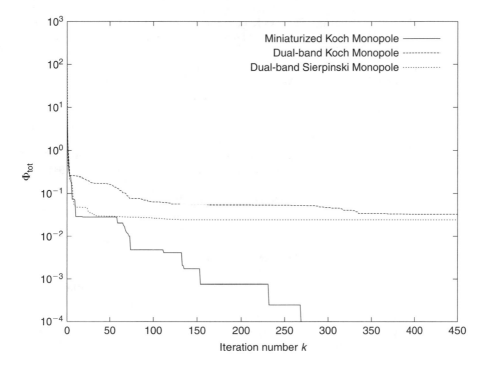

FIGURE 2.13
Behavior of the cost function versus the iteration number registered during the synthesis of the prefractal WiMAX antennas.

PSO-based optimization procedure. The design process as well as the resulting multiband antenna prototypes have been validated through experimental and numerical tests. The obtained results in terms of both gains and VSWR values have confirmed the effectiveness of the proposed design procedure as well as its feasibility for WiMAX applications.

References

1. J. D. Kraus, *Antennas*. New York: McGraw-Hill, 1988.
2. C. A. Balanis, *Antenna Theory: Analysis and Design*. New York: Wiley, 1996.
3. W. L. Stutzman and G. A. Thiele, *Antenna Theory and Design*. New York: Wiley, 1981.
4. S.-W. Su and K.-L. Wong, Wideband antenna integrated in a system in package for WLAN/WiMAX operation in a mobile device, *Microw. Opt. Technol. Lett.*, vol. 48, no. 10, pp. 2048–2053, Oct. 2006.
5. S.-W. Su, K.-L. Wong, C.-L. Tang, and S.-H.-H. Yeh, Wideband monopole antenna integrated within the front-end module package, *IEEE Trans. Antennas Propagat.*, vol. 54, pp. 1888–1891, Jun. 2006.

6. J. M. Ide, S. P. Kingsley, S. G. O'Keefe, and S. A. Saario, A novel wide band antenna for WLAN applications, in *Proc. IEEE AP-S Int. Symp.*, vol. 4, 2005, pp. 243–246.
7. G. R. DeJean, T. T. Thai, and M. M. Tentzeris, Design of microstrip bi-Yagi and microstrip quad-Yagi antenna arrays for WLAN and millimeter-wave applications, in *Proc. IEEE AP-S Int. Symp.*, 2006, pp. 989–992.
8. C.-I. Lin, K.-L. Wong, and S.-H. Yeh, Wideband EMC chip antenna for WLAN/WiMAX operation in the sliding mobile phone, *Microw. Opt. Tech. Lett.*, vol. 48, no. 7, pp. 1362–1366, Jul. 2006.
9. K.-L. Wong, C.-H. Wu, and F.-S. Chang, Broadband coaxial antenna for WiMAX access-point application, *Microw. Opt. Tech. Lett.*, vol. 48, no. 4, pp. 641–644, Apr. 2006.
10. Y.-T. Liu and K.-L. Wong, A wideband stubby monopole antenna and a GPS antenna for WiMAX mobile phones with E911 function, *Microw. Opt. Tech. Lett.*, vol. 46, no. 5, pp. 485–487, Sep. 2005.
11. K.-L. Wong, W.-C. Su, and F.-S. Chang, Wideband internal folded planar monopole antenna for UMTS/WiMAX folder-type mobile phone, *Microw. Opt. Tech. Lett.*, vol. 48, no. 2, pp. 324–327, Feb. 2006.
12. K.-L. Wong and L.-C. Chou, Internal composite monopole antenna for WLAN/WiMAX operation in a laptop computer, *Microw. Opt. Tech. Lett.*, vol. 48, no. 5, pp. 868–871, May 2006.
13. W.-J. Liao, Y.-C. Lu, and H.-T. Chou, A novel multiband dipole antenna with a microstrip loop feed, *Microw. Opt. Tech. Lett.*, vol. 49, no. 1, pp. 237–241, Jan. 2007.
14. J. M. Floc'h and H. Rmili, Design of multiband printed dipole antennas using parasitic elements, *Microw. Opt. Tech. Lett.*, vol. 48, no. 8, pp. 1639–1645, Aug. 2006.
15. W.-C. Su and K.-L. Wong, Internal PIFAS for UMTS/WLAN/WiMAX multi-network operation for a USB dongle, *Microw. Opt. Tech. Lett.*, vol. 48, no. 11, pp. 2249–2253, Nov. 2006.
16. B. B. Mandelbrot, *The Fractal Geometry of Nature*. New York: W. H. Freeman and Company, 1983.
17. H. O. Peitgen, H. Jurgens, and D. Saupe, *Chaos and Fractals, New Frontiers of Science*. New York: Springer, 1990.
18. D. H. Werner and R. Mittra, *Frontiers in Electromagnetics*. Piscataway: IEEE Press, 2000.
19. J. Gianvittorio and Y. Rahmat-Samii, Fractals antennas: A novel antenna miniaturization technique, and applications, *IEEE Antennas Propagat. Mag.*, vol. 44, pp. 20–36, Feb. 2002.
20. S. R. Best, A comparison of the resonant properties of small space-filling fractal antennas, *IEEE Antennas Wireless Propagat. Lett.*, vol. 2, pp. 197–200, 2003.
21. J. M. Gonzàles-Arbesù, S. Blanch, and J. Romeu, Are space-filling curves efficient small antennas? *IEEE Antennas Wireless Propagat. Lett.*, vol. 2, pp. 147–150, 2003.
22. J. M. Gonzàles-Arbesù and J. Romeu, On the influence of fractal dimension on radiation efficiency and quality factor of self-resonant prefractal wire monopoles, in *Proc. IEEE Antennas Propagat. Symp.*, Jun. 2003.
23. C. P. Baliarda, J. Romeu, and A. Cardama, The Koch monopole: A small fractal antenna, *IEEE Antennas Propagat. Mag.*, vol. 48, pp. 1773–1781, Nov. 2000.

24. S. R. Best, On the performance properties of the Koch fractal and other bent wire monopoles, *IEEE Trans. Antennas Propagat.*, vol. 51, pp. 1292–1300, Jun. 2003.
25. A. W. Rudge, K. Milne, A. D. Olver, and P. Knight, *The Handbook of Antenna Design.* London, UK: Peter Peregrinus, 1986.
26. Y. T. Lo and S. W. Lee, *Antenna Handbook.* New York, NY: Van Nostrand Reinhold, 1988.
27. D. H. Werner, P. L. Werner, and K. H. Church, Genetically engineered multiband fractal antennas, *Electron. Lett.*, vol. 37, pp. 1150–1151, Sep. 2001.
28. M. F. Pantoja, F. G. Ruiz, A. R. Bretones, R. G. Martin, J. M. Gonzales-Arbesù, J. Romeu, and J. M. Rius, GA design of wire pre-fractal antennas and comparison with other euclidean geometries, *IEEE Antennas Wireless Propagat. Lett.*, vol. 2, pp. 238–241, 2003.
29. M. F. Pantoja, F. G. Ruiz, A. R. Bretones, S. G. Garcia, R. G. Martin, J. M. Gonzales-Arbesù, J. Romeu, J. M. Rius, P. L. Werner, and D. H. Werner, GA design of small thin-wire antennas: Comparison with Sierpinski type prefractal antennas, *IEEE Trans. Antennas Propagat.*, vol. 54, pp. 1879–1882, Jun. 2006.
30. C. Puente, J. Romeu, R. Bartoleme, and R. Pous, Perturbation of the Sierpinski antenna to allocate operating bands, *Electron. Lett.*, vol. 32, pp. 2186–2188, Nov. 1996.
31. R. Azaro, F. De Natale, M. Donelli, E. Zeni, and A. Massa, Synthesis of a prefractal dual-band monopolar antenna for GPS applications, *IEEE Antennas Wireless Propagat. Lett.*, vol. 5, pp. 361–364, 2006.
32. R. Azaro, E. Zeni, M. Zambelli, and A. Massa, Synthesis and optimization of pre-fractal multiband antennas, in *Proc. EuCAP 2006*, Nice, France, 6–10 Nov. 2006.
33. R. Azaro, M. Donelli, D. Franceschini, E. Zeni, and A. Massa, Optimized synthesis of a miniaturized SARSAT band pre-fractal antenna, *Microw. Opt. Tech. Lett.*, vol. 48, no. 11, pp. 2205 2207, Nov. 2006.
34. J. R. Robinson and Y. Rahmat-Samii, Particle swarm optimization in electromagnetics, *IEEE Trans. Antennas Propagat.*, vol. 52, pp. 771–778, Feb. 2004.
35. M. Donelli and A. Massa, Computational approach based on a particle swarm optimizer for microwave imaging of two-dimensional dielectric scatterers, *IEEE Trans. Microwave Theory Tech.*, vol. 53, pp. 1761–1776, May 2005.
36. R. Azaro, F. De Natale, M. Donelli, A. Massa, and E. Zeni, Optimized design of a multi-function/multi-band antenna for automotive rescue systems, *IEEE Trans. Antennas Propagat.*—Special Issue on *Multifunction Antennas and Antenna Systems*, vol. 54, pp. 392–400, Feb. 2006.
37. R. F. Harrington, *Field Computation by Moment Methods.* Malabar, FL: Robert E. Krieger Publishing Co., 1987.
38. J. Kennedy, R. C. Eberhart, and Y. Shi, *Swarm Intelligence.* San Francisco: Morgan Kaufmann Publishers, 2001.
39. Y. Rahmat-Samii, Frontiers in evolutionary optimization techniques applied to antenna designs: Genetic algorithms and particle swarm optimization, in *Proc. 13th Int. Symp. Antennas (JINA 2004)*, Nice, France, 8–10 Nov. 2004.
40. D. W. Boringer and Douglas H. Werner, Particle swarm optimization versus genetic algorithms for phased array synthesis, *IEEE Trans. Antennas Propagat.*, vol. 52, pp. 771–779, Mar. 2004.
41. M. Donelli, R. Azaro, F. G. B. De Natale, and A. Massa, An innovative computational approach based on a particle swarm strategy for adaptive phased-arrays control, *IEEE Trans. Antennas Propagat.*, vol. 54, pp. 888–898, Mar. 2006.

3

Space–Time Coding and Application in WiMAX

Naofal Al-Dhahir, Robert Calderbank, Jimmy Chui,
Sushanta Das, and Suhas Diggavi

CONTENTS

3.1   Introduction

Next-generation wireless systems aim to support both voice and high capacity flexible data services with limited bandwidth. Multiplicative and additive distortions inherent to the wireless medium make this difficult, and extensive research efforts are focused on developing efficient technologies to support reliable wireless communications. One such technology is multiple-input multiple-output (MIMO) communication systems with multiple antennas at both the transmitter and the receiver. Information-theoretic analysis in Refs. 1 and 2 shows that multiple antennas at the transmitter and the receiver enable very high data rates. Another key technology enabling high-rate communications over the wireless channels is orthogonal frequency division multiplexing (OFDM) [3]. In this chapter, we address and investigate a special class of MIMO, namely, space–time block codes (STBCs), and its application in WiMAX, the next-generation OFDM system based on IEEE 802.16 standard [4,5].

The main attribute that dictates the performance of wireless communications is uncertainty (randomness). Randomness exists in the users' transmission channels, as well as in the users' geographical locations. The spatial separation of antennas results in additional randomness. Space–time codes, introduced in Ref. 6, improve the reliability of communication over fading channels by correlating the transmit signals in both spatial and temporal dimensions, thus attaining *diversity*. We broadly define diversity as the method of conveying information through multiple independent instantiations of these random attenuations. The inherent undesirable trait of randomness is used as the foundation to enhance performance!

Space–time coding has received considerable attention in both academic and industrial circles:

- First, it improves the downlink performance without the need for multiple receive antennas at the terminals. For example, space–time coding techniques in wideband CDMA achieve substantial capacity gains owing to the resulting smoother fading which, in turn, makes power control more effective and reduces the transmitted power [7].
- Second, it can be elegantly combined with channel coding, as shown in Ref. 6, which realizes a coding gain in addition to the spatial diversity gain.
- Third, it does not require channel state information (CSI) at the transmitter, i.e., it operates in open-loop mode. Thus, it eliminates the need for an expensive and, in case of rapid channel fading, unreliable reverse link.

- Finally, it has been shown to be robust against nonideal operating conditions, such as antenna correlation, channel estimation errors, and Doppler effects [8,9].

An elegant and simple subclass of space–time codes are STBCs, which are able to provide high information rate to serve a large number of users over a wide coverage area with adequate reliability. This chapter describes the principal codes in this class that appear in the IEEE 802.16-2004 standard [4] and its 802.16e-2005 amendment [5]. We also examine other prospective candidates, including a novel nonlinear design based on quaternions.

The WiMAX Forum [10], the associated industry consortium of IEEE Std 802.16, promises to deliver broadband connectivity with a data rate up to 40 Mbps and a coverage radius of 3–10 km using multiple-antenna technology. The IEEE 802.16-2004 standard is designed for stationary transmission, and the 802.16e amendment deals with both stationary and mobile transmissions. The standard and the amendment define the physical (PHY) layer specifications used in WiMAX. We identify two important features of WiMAX: (i) it uses OFDM that has an inherent robustness against multipath propagation and frequency-selective channels and (ii) it allows the choice of multiple-antenna systems, e.g., MIMO systems and adaptive antenna systems (AASs). Multiple-antenna systems can be implemented very easily with OFDM. The AAS transmits multiple spatially separated overlapped signals using space-division multiple access and MIMO often uses spatial multiplexing (e.g., BLAST). Even though both of these schemes are modeled as high data rate providers in WiMAX, the simplest example of multiple-antenna systems is the well-known Alamouti Code [11]. The use of multiple antennas at the transmitter and the receiver enhances the system spectral efficiency, supports better error rate, and increases coverage area. These benefits come at no extra cost of bandwidth and power.

We demonstrate the value of using multiple antennas and STBCs in WiMAX by examining the performance gain of our nonlinear quaternionic code, which utilizes four transmit antennas and achieves full diversity, and comparing it with a single-input single-output (SISO) implementation. The gain in signal-to-noise ratio (SNR), achieved through the use of this novel 4×4 STBC over SISO in a WiMAX environment, translates to a 50% increase in the cell coverage area assuming only one receive antenna. By adding a second receive antenna, the percentage increase becomes 166%.

In the following sections, we will review the details of multiple-antenna transmission schemes and the design criteria for space–time codes. We also present codes that appear in the standard, as well as other notable codes, and relate them to the underlying theory. We examine how STBCs are used in practice, including implementation in OFDM and nonideal considerations. At the end of this chapter, we present our rate-1 full-diversity code for four transmit antennas, and demonstrate the value of this STBC in the WiMAX environment.

3.2 Space–Time Codes: A Primer

3.2.1 System Model: Quasi-Static Rayleigh Fading Channel

WiMAX is a broadband transmission system, and thus inherently has inter-symbol interference (ISI) due to the frequency-selective nature of the wireless channel. The use of OFDM in WiMAX divides the entire bandwidth into many parallel narrowband channels, each with flat-fading characteristics. In this section, we focus our discussion on the theory of space–time coding on flat-fading channels. Our motivation for this approach is to emulate the design and application of space–time coding in WiMAX. At the end of this section, we examine the ISI channel in more detail and discuss its implications on more optimal designs for space–time codes.

The challenge of communication over Rayleigh fading channels is that the error probability decays only inversely with SNR, compared with the exponential decay observed on AWGN channels. Space–time coding [6] is a simple method that enhances reliability by increasing the decay of error probability through diversity.

There has been extensive work on the design of space–time codes since their introduction. In this section, we describe the basic design principles of space–time codes. We define the notions of rate and diversity, and examine the governing trade-off law between the two.

We formulate the system model as follows. The channel model is a MIMO quasi-static Rayleigh flat-fading channel with M_t transmit antennas and M_r receive antennas. The quasi-static assumption indicates that the channel gain coefficients remain constant for the duration of the codeword and change independently for each instantiation. The flat-fading assumption allows each transmitted symbol to be represented by a single tap in the discrete-time model with no ISI. We assume independent Rayleigh coefficients, i.e., each fading coefficient is an i.i.d. circular-complex normal random variable $\mathcal{CN}(0, 1)$. White Gaussian noise is also added at the receiver. The system model also assumes that the receiver has perfect CSI, whereas the transmitter does not have any CSI.

This system model can be described in matrix notation. The transmitted (received) codewords are represented by matrices, where the rows are indexed by the transmit (receive) antennas, the columns are indexed by time slots in a data frame, and the entries are the transmitted (received) symbols in baseband. For each codeword transmission,

$$\mathbf{R} = \mathbf{H}\mathbf{X} + \mathbf{Z} \tag{3.1}$$

where $\mathbf{R} \in \mathbf{C}^{M_r \times T}$ is the received signal matrix, $\mathbf{H} \in \mathbf{C}^{M_r \times M_t}$ the quasi-static channel matrix representing the channel gains of the $M_r M_t$ paths from each transmit antenna to each receive antenna, $\mathbf{X} \in \mathbf{C}^{M_t \times T}$ the transmitted codeword, and $\mathbf{Z} \in \mathbf{C}^{M_r \times T}$ the additive white Gaussian noise where each entry has distribution $\mathcal{CN}(0, \sigma^2)$. There is also a power constraint, where the transmitted symbols have an average power of P, where the average is taken across all codewords over both spatial and temporal components. We assume that $P = 1$.

In this context, the length T of a codeword is a design parameter. In practice, T must be less than the coherence interval of the channel, to satisfy the quasi-static assumption.

3.2.2 Diversity Gain and Coding Gain

For a SISO channel, Equation 3.1 simplifies to

$$\mathbf{r}[m] = h\mathbf{x}[m] + \mathbf{z}[m]$$

We use lowercase variables to emphasize the vector nature in the SISO environment. The instantaneous received SNR is the product $|h|^2$SNR. If $|h|^2$SNR $\gg 1$ then the separation between signal points is significantly larger than the standard deviation of the Gaussian noise, and error probability is very small since the tail of the Q-function decays rapidly. On the contrary, if $|h|^2$SNR $\ll 1$ then the separation between signal points is much less than the standard deviation of the noise, and the error probability is significant. Error events in the high SNR regime most often occur because the channel is in a deep fade ($|h|^2$SNR < 1), and not as a result of high additive noise. For the Rayleigh fading channel, the probability of error (for each bit and for the codeword) is proportional to $1/$SNR at sufficiently high SNR.

The independent Rayleigh fading encountered across different codewords can be exploited to provide *diversity*. By repeating the same codeword M times, the probability of error will be proportional to $1/$SNR^M. Reliable communication for a particular codeword is possible when at least one of the M transmissions encounter favorable conditions, that is, does not see a deep fade. A supplementary outer code can help achieve better performance through additional coding gain.

Diversity is also introduced through multiple antennas at the transmitter. During the transmission of a single codeword, there are $M_t M_r$ observed independent fades. Space–time codes correlate the transmitted symbols in a codeword, which protect each (uncoded) symbol up to $M_t M_r$ times for each independent fade it encounters. If at least one path is strong, reliable communication for that symbol is possible. Diversity can also occur at the receiver if it is equipped with multiple antennas, provided that the receive antennas are spaced sufficiently far enough so that the correlation of the fade coefficients is negligible.

So far, we have qualitatively discussed diversity as the notion of sending correlated symbols across multiple paths from the transmitter to the receiver. Diversity is quantified through the notion of diversity gain or diversity order, which represents the decay of error probability in the high SNR regime.

DEFINITION 3.1   A coding scheme with an average error probability $\mathbb{P}_e$(SNR) *as a function of* SNR *that behaves as*

$$\lim_{\text{SNR} \to \infty} \frac{\log[\mathbb{P}_e(\text{SNR})]}{\log(\text{SNR})} = -d \tag{3.2}$$

is said to have a diversity gain of d.

In words, a scheme with diversity order d has an error probability at high SNR behaving as* $\mathbb{P}_e(\text{SNR}) \doteq \text{SNR}^{-d}$.

One can approximate the performance of a space–time code by determining the worst pairwise error probability (PEP) between two candidate codewords. This leads to the rank criterion for determining the diversity order of a space–time code [6,12]. The PEP between two codewords $\mathbf{X}_i$ and $\mathbf{X}_j$ can be determined by properties of the difference matrix $\Delta(\mathbf{X_i}, \mathbf{X_j}) = \mathbf{X}_i - \mathbf{X}_j$. When there is no ambiguity we will denote the difference simply by Δ.

$$\mathbb{P}(\mathbf{X}_i \to \mathbf{X}_j) = \mathbb{E}_{\mathbf{H}}[\mathbb{P}(\mathbf{X}_i \to \mathbf{X}_j \mid \mathbf{H})]$$

$$= \mathbb{E}_{\mathbf{H}}\left[Q\left(\frac{||\mathbf{H}\Delta||}{\sqrt{2N_0}}\right)\right] \tag{3.3}$$

Under the Rayleigh fading assumption, Equation 3.3 can be bounded above using the Chernoff bound, as in Ref. 6. If q is the rank of Δ and $\{\lambda_n\}_{n=1}^{q}$ are the nonzero eigenvalues of $\Delta\Delta^*$, then the upper bound is given by

$$\mathbb{P}(\mathbf{X}_i \to \mathbf{X}_j) \le \left(\prod_{n=1}^{q} \lambda_n\right)^{-M_r} \left(\frac{\text{SNR}}{4}\right)^{-qM_r} \tag{3.4}$$

It can be shown that the exact expression for the asymptotic PEP at high SNR is a multiplicative constant (dependent only on M_r and q) of the upper bound given in Equation 3.3 [13].

For a space–time code with a fixed-rate codebook $\mathcal{C}$, the worst PEP corresponds to the pair of codewords for which the difference matrix has the lowest rank and the lowest product measure (the product of the nonzero eigenvalues). By a simple union-bound argument, it follows that the diversity gain d is given by

$$d = M_r \min_{\mathbf{X}_i \neq \mathbf{X}_j \in \mathcal{C}} \text{rank}[\Delta(\mathbf{X}_i, \mathbf{X}_j)] \tag{3.5}$$

The expression in Equation 3.4 also leads one to the following definition of coding gain.

---

* We use the notation $\doteq$ to denote exponential equality, i.e., $g(\text{SNR}) \doteq \text{SNR}^a$ means that $\lim_{\text{SNR} \to \infty} \log g(\text{SNR}) / \log \text{SNR} = a$. Moreover, if $g(\text{SNR}) \doteq f(\text{SNR})$, it means that $\lim_{\text{SNR} \to \infty} \log g(\text{SNR}) / \log \text{SNR} = \lim_{\text{SNR} \to \infty} \log f(\text{SNR}) / \log \text{SNR}$.

DEFINITION 3.2   *The coding gain for a space–time code with a fixed-rate codebook is given by the quantity*

$$\left(\prod_{n=1}^{q} \lambda_n\right)^{1/q} \tag{3.6}$$

that corresponds to the worst PEP.

Both the coding gain and the diversity order contribute to the error probability. Hence, two criteria for code design, as indicated in Ref. 6, are to design the codebook $\mathcal{C}$ such that the following are satisfied.

Rank criterion: Maximize the minimum rank of the difference $\mathbf{X}_i - \mathbf{X}_j$ over all distinct pairs of space–time codewords $\mathbf{X}_i, \mathbf{X}_j$.

Determinant criterion: For a given diversity d, maximize the minimum product of the nonzero singular values of the difference $\mathbf{X}_i - \mathbf{X}_j$ over all distinct pairs of space–time codewords $\mathbf{X}_i, \mathbf{X}_j$ whose difference has rank d.

We note that these criteria optimize the code for the high SNR regime. Optimizing for lower SNR values implies using the Euclidean distance metric [14]. Recent results also suggest that examining the effect of multiple interactions between codewords is necessary for a more accurate comparison between codes [15].

3.2.3   Trade-Offs between Diversity and Rate

A natural question that arises is to determine how many codewords can we have, which allow us to attain a certain diversity order. One point of view is to fix the constellation, and examine the trade-off between the number of codewords and the code's diversity gain. A second is to allow the constellation to grow as SNR increases. This latter viewpoint is motivated because the capacity of the multiple-antenna channel grows with SNR behaving as $\min(M_r, M_t) \log(\text{SNR})$ [1,16], at high SNR even for finite M_r, M_t.

3.2.3.1   Trade-Off for Fixed Constellations

For the quasi-static Rayleigh flat-fading channel, this has been examined in Ref. 6 where the following result was obtained.

THEOREM 3.1 [6]
For a static constellation per transmitted symbol, if the diversity order of the system is qM_r, then the rate R that can be achieved (in terms of symbols per symbol-time period) is bounded as

$$R \leq M_t - q + 1 \tag{3.7}$$

3.2.3.2 Diversity-Multiplexing Trade-Off

Zheng and Tse [16] define a multiplexing gain of a transmission scheme as follows.

DEFINITION 3.3 *A coding scheme that has a transmission rate of R(SNR) as a function of* SNR *is said to have a multiplexing gain r if*

$$\lim_{\text{SNR}\to\infty} \frac{R(\text{SNR})}{\log(\text{SNR})} = r \tag{3.8}$$

Therefore, the system has a rate of $r\log(\text{SNR})$ at high SNR. The main result in Ref. 16 states that

THEOREM 3.2 [16]
For $T \geq M_t + M_r - 1$, and $K = \min(M_t, M_r)$, the optimal trade-off curve $d^*(r)$ is given by the piece-wise linear function connecting points in $[k, d^*(k)], k = 0, \dots, K$ where

$$d^*(k) = (M_r - k)(M_t - k) \tag{3.9}$$

Both Theorems 3.1 and 3.2 show the tension between achieving high-rate and high-diversity. If $r = k$ is an integer, the result can be interpreted as using $M_r - k$ receive antennas and $M_t - k$ transmit antennas to provide diversity while using k antennas to provide the multiplexing gain. Clearly, this result means that one can get large rates that grow with SNR if we reduce the diversity order from the maximum achievable. This diversity-multiplexing trade-off implies that a high multiplexing gain comes at the price of decreased diversity gain and is associated with a trade-off between error probability and rate.

This tension between rate and diversity (reliability) demonstrates that different codes will be suitable for different situations. The code choice is by design, and can be influenced by factors such as quality of service and maximum tolerable delay. Section 3.3 demonstrates a selection of various space–time transmission schemes, including those that appear in the IEEE 802.16 standards. These codes range from those that maximize spatial multiplexing (V-BLAST) to codes with full-diversity gain (the Alamouti Code, the Golden Code, and our code based on quaternions).

3.2.4 The ISI Channel

We have justified our use of the idealized flat-fading channel due to the narrowband characteristics for each tone in an OFDM system. This allows for simple code design criteria, as each tone is coded independent of the others. The theory of space–time codes is not limited to flat-fading, however, and can be extended to ISI channels such as frequency-selective channels or channels with multipath. The general model for ISI channels is almost identical to Equation 3.1 except the model now includes a new parameter v, the number of taps required to characterize a given ISI channel.

The rank and determinant criteria for the ISI channel are similar to those given in Section 3.2.2. The main problem in practice is to construct such codes that do not have large decoding complexity. The trade-off between performance and complexity is more prominent for ISI code design. A thorough examination can be found in Ref. 17.

The parameter v allows an increase in the maximum diversity gain for the system by a factor of v. Intuitively, this is a result of using frequency as an additional means of diversity for a frequency-selective channel. As such, it affects the tension between rate and diversity. For example, the maximum achievable rate for fixed-rate codes given a diversity order of qM_r is

$$R \leq M_t - (q/v) + 1 \tag{3.10}$$

The diversity-multiplexing trade-off for the SISO ISI channel is [18]

$$d^*(r) = v(1 - r) \tag{3.11}$$

For a SISO OFDM system with N tones, if there are v i.i.d. ISI taps, then the frequency domain taps separated by N/v are going to be independent. For simplicity, we assume that all tones are in use and that N/v is an integer. Therefore, we have N/v sets of v parallel channels, where within each set the frequency-domain fading are independent. Then by using a diversity-multiplexing optimal code for the parallel fading channel for each of these sets, we can achieve Equation 3.11. Trivially, for one degree of freedom, i.e., $\min(M_t, M_r) = 1$, the SISO result can be extended to $d^*(r) = v \max(M_t, M_r) \times (1 - r)$, and the same coding architecture applies.

In WiMAX, frequency diversity is achieved by using an outer code and frequency interleaving (e.g., Ref. 4, Sections 8.3.3 and 8.4.9). The use of this outer code, along with STBCs designed for the flat-fading channel, achieves a very good improvement in performance, which we will demonstrate in Section 3.6.

3.3 Space–Time Block Codes

3.3.1 Spatial Multiplexing

One strategy for constructing codes is to maximize rate and achieve the greatest spectral efficiency. Spatial multiplexing achieves this goal by transmitting uncorrelated data over space and time, and relies on a sophisticated decoding mechanism to separate the data streams at the receiver. Strictly speaking, spatial multiplexing is not a method of space–time coding; it does not provide transmit diversity. However, it can be considered an extreme case in the diversity-multiplexing trade-off curve; it trades off diversity for maximal rate. The performance of these codes is highly dependent on the decoding algorithm at the receiver and the instantaneous channel characteristics.

Codes exploiting spatial multiplexing occur in the standard for two transmit antennas (Section 8.4.8.3.3, Code B), three transmit antennas (Section 8.4.8.3.4, Code C), and four transmit antennas (Section 8.4.8.3.5, Code C). The symbols may support different code rates.

$$\begin{pmatrix} x_1 \\ x_2 \end{pmatrix}; \qquad \begin{pmatrix} x_1 \\ x_2 \\ x_3 \end{pmatrix}; \qquad \begin{pmatrix} x_1 \\ x_2 \\ x_3 \\ x_4 \end{pmatrix}$$

These constructions fall under the Bell Labs Layered Space–Time Architecture (BLAST) framework [1,19], as V-BLAST codes. A major challenge in realizing this significant additional throughput gain in practice is the development of cost-effective low-complexity and highly optimal receivers. The receiver signal-processing functions are similar to a decision feedback equalizer operating in the spatial domain where the nulling operation is performed by the feedforward filter and the interference cancellation operation is performed by the feedback filter [20]. As with all feedback-based detection schemes, V-BLAST suffers from error propagation effects.

3.3.2 The Alamouti Code

The Alamouti Code [11] was discovered as a method to provide transmit diversity in the same manner maximum-ratio receive combining provides receive diversity. This is achieved by correlating the transmit symbols spatially across two transmit antennas, and temporally across two consecutive time intervals. In the notation of Equation 3.1, the Alamouti Code encodes two symbols x_1 and x_2 as the following 2×2 matrix:

$$(x_1, x_2) \rightarrow \begin{pmatrix} x_1 & x_2 \\ -x_2^* & x_1^* \end{pmatrix} \tag{3.12}$$

Unlike spatial multiplexing codes, the Alamouti Code achieves the same rate as SISO but attains maximum diversity gain for two transmit antennas. Achieving diversity is a second strategy for designing space–time codes, and it comes at the expense of reduced rate.

The coding and decoding mechanisms for the Alamouti Code are remarkably simple, and equally effective. It is not surprising to see the appearance of the Alamouti Code many times in the standard (e.g., Section 8.3.8.2 in Ref. 4, Section 8.4.8.3.3 in Ref. 5, and variants for three and four antennas described later in this section). Next, we demonstrate the simplicity of Alamouti Code decoding for the case of one receive antenna.

The receive antenna obtains the signals r_1, r_2 over the two consecutive time slots for the corresponding codeword. They are given by

$$\begin{pmatrix} r_1 \\ -r_2^* \end{pmatrix} = \begin{pmatrix} h_1 & h_2 \\ -h_2^* & h_1^* \end{pmatrix} \begin{pmatrix} x_1 \\ -x_2^* \end{pmatrix} + \begin{pmatrix} z_1 \\ -z_2^* \end{pmatrix} \tag{3.13}$$

where h_1, h_2 are the path gains from the two transmit antennas to the mobile, and the noise samples z_1, z_2 are independent samples of a zero-mean complex Gaussian random variable with noise energy N_0 per complex dimension. Thus

$$\mathbf{r} = \mathbf{H}\mathbf{s} \mid \mathbf{z} \tag{3.14}$$

where the matrix $\mathbf{H}$ is orthogonal. If the receiver knows the path gains h_1 and h_2, then it is able to form

$$\mathbf{H}^*\mathbf{r} = ||\mathbf{h}||^2\mathbf{s} + \mathbf{z}' \tag{3.15}$$

where the new noise term $\mathbf{z}'$ remains white. This allows the *linear-complexity* maximum likelihood (ML) decoding of x_1, x_2 to be done independently rather than jointly.

We note that the Alamouti Code provides the same rate as an equivalent SISO channel. Using the Alamouti Code provides us with higher reliability, owing to the increased diversity.

The Alamouti Code is defined for only two transmit antennas. In the standard, variants using three and four transmit antennas are described as well. These codes are constructed by using only two transmit antennas over two consecutive time intervals, leaving the other antenna(s) effectively in an off state.

To increase rate, the off states can be removed and replaced by extra symbols. For three transmit antennas, these extra symbols are transmitted in an uncorrelated fashion; for four transmit antennas, the extra symbols are sent in a correlated manner by means of the Alamouti Code. Note that in each scenario, each unique symbol uses the same transmitted power. These may be considered as a combination of spatial multiplexing in conjunction with Alamouti codewords.

Examples of three-antenna Alamouti-based space–time codes are listed in Ref. 5, Section 8.4.8.3.4. Codes A_1 and B_1 have the following form:

$$A_1 = \begin{pmatrix} x_1 & x_2 & 0 & 0 \\ -x_2^* & x_1^* & x_3 & x_4 \\ 0 & 0 & -x_4^* & x_3^* \end{pmatrix} \quad B_1 = \begin{pmatrix} \sqrt{\frac{3}{4}} & 0 & 0 \\ 0 & \sqrt{\frac{3}{4}} & 0 \\ 0 & 0 & \sqrt{\frac{3}{2}} \end{pmatrix} \begin{pmatrix} x_1 & x_2 & x_3 & x_4 \\ -x_2^* & x_1^* & -x_4^* & x_3^* \\ x_5 & x_6 & x_7 & x_8 \end{pmatrix}$$

The four-antenna Alamouti variants are similar and can be found in Ref. 5, Section 8.4.8.3.5. Two of the codes are provided below. We note that the decoding for Code B can be achieved by successive cancellation [21,22].

$$
A = \begin{pmatrix} x_1 & x_2 & 0 & 0 \\ -x_2^* & x_1^* & 0 & 0 \\ 0 & 0 & x_3 & x_4 \\ 0 & 0 & -x_4^* & x_3^* \end{pmatrix} \qquad B = \begin{pmatrix} x_1 & x_2 & x_5 & x_6 \\ -x_2^* & x_1^* & x_7 & x_8 \\ x_3 & x_4 & -x_6^* & x_5^* \\ -x_4^* & x_3^* & -x_8^* & x_7^* \end{pmatrix}
$$

3.3.3 The Golden Code

Of the above codes, one provides diversity (Alamouti) while the other provides rate (spatial multiplexing). Can we obtain a code that achieves the diversity order of Alamouti and the rate of spatial multiplexing? According to the diversity multiplexing trade-off (Theorem 3.2), it appears that such a code may exist. The answer is in the affirmative, and is given by the Golden Code.

The Golden Code is a space–time code for a system with two transmit antennas. It has the property that four complex symbols can be encoded over two time slots, yet achieves full diversity. This code was independently discovered in Refs. 23 and 24.

The Golden Code encodes four QAM symbols x_1, x_2, x_3, x_4 to the following 2×2 matrix:

$$
(x_1, x_2, x_3, x_4) \rightarrow \frac{1}{\sqrt{5}} \begin{pmatrix} \alpha(x_1 + \theta x_2) & \alpha(x_3 + \theta x_4) \\ i\bar{\alpha}(x_3 + \bar{\theta} x_4) & \bar{\alpha}(x_1 + \bar{\theta} x_2) \end{pmatrix} \tag{3.16}
$$

where

$$
\theta = \frac{1 + \sqrt{5}}{2} \qquad\qquad \alpha = 1 + i - i\theta
$$

$$
\bar{\theta} = \frac{1 - \sqrt{5}}{2} = 1 - \theta \qquad \bar{\alpha} = 1 + i - i\bar{\theta}
$$

In Ref. 25, it is shown that nonzero matrices of this form (and hence any nonzero difference matrix) have nonvanishing determinant. That is to say, the set of all determinants can never be arbitrarily made close to 0. Indeed, the minimum absolute value for the determinant of nonzero matrices of this code is 0.2. Decoding of the Golden Code can be done with sphere decoding [26–28].

It can be shown that the Alamouti Code, as well as the spatial multiplexing code, does not achieve the diversity-multiplexing trade-off, whereas the Golden Code does [23]. This appealing property is a reason for its appearance in the standard (Section 8.4.8.3.3, Code C in Ref. 5). The code in the standard is a variation of the Golden Code (Equation 3.16) [29].

3.3.4 Other Space–Time Block Codes

In the remainder of this section, we examine other codes that do not make an appearance in the standard.

Two attractive properties of the Alamouti STBC are the ability to separate symbols at the receiver with linear processing and the achievement of maximal diversity gain. These properties define the broader class of STBCs, namely orthogonal space–time block codes [30]. Orthogonal designs achieve maximal diversity at a linear (in constellation size) decoding complexity. Tarokh et al. prove that the only full-rate complex orthogonal design only exists for the case of two transmit antennas, and is given by the Alamouti Code. As the number of transmit antennas increases, the available rate becomes unattractive [31].

For four transmit antennas, the maximum rate is 3/4, and one such example is presented in Ref. 30:

$$\begin{bmatrix} x_1 & x_2 & x_3 & 0 \\ -\bar{x}_2 & \bar{x}_1 & 0 & x_3 \\ \bar{x}_3 & 0 & -\bar{x}_1 & x_2 \\ 0 & \bar{x}_3 & -\bar{x}_2 & -x_1 \end{bmatrix} \tag{3.17}$$

Given that full-rate orthogonal designs are limited in number, it is natural to relax the requirement that linear processing at the receiver be able to separate all transmitted symbols. The lack of a full-rate complex design for even four transmit antennas motivated Jafarkhani [32] to consider the quasi-orthogonal space–time block code

$$\mathbf{S} = \left[\begin{array}{cc|cc} x_1 & x_2 & x_3 & x_4 \\ -x_2^* & x_1^* & -x_4^* & x_3^* \\ \hline -x_3^* & -x_4^* & x_1^* & x_2^* \\ x_4 & -x_3 & -x_2 & x_1 \end{array} \right]$$

where the structure of each 2×2 block is identical to that of the Alamouti Code. ML decoding can be achieved by processing *pairs* of symbols, namely (x_1, x_4) and (x_2, x_3) for this code.

This construction only achieves a diversity gain of two for every receive antenna. Full diversity can be achieved at the expense of signal constellation expansion, for example, by rotating the symbols x_1 and x_3 (see e.g., Ref. 33). The optimal angle of rotation depends on the base constellation.

There exist many other space–time codes in literature, including trellis-based constructions (including super-orthogonal codes [34]), linear dispersion codes [35], layered space–time coding [36], threaded algebraic space–time codes [37], and more recently, perfect STBCs [38]. The Golden Code can be considered as an instance of the last two constructions.

3.4   Application of Space–Time Coding in WiMAX

3.4.1   Space–Time Coding in OFDM

STBCs based on the flat-fading design has a straightforward implementation in OFDM systems. The modulated symbols are encoded into space–time codes before the IFFT operation. Figure 3.1 depicts the implementation. Like the SISO case, multipath along each transmitter–receiver path is mitigated by use of the cyclic prefix. The only additional condition is that the coherence time of the physical channel must exceed that of T times the duration of an OFDM symbol, where T is the length of the codeword. During the span of T OFDMA symbols, each subcarrier in use transmits one space–time code.

3.4.2   Channel Estimation

Our assumption of perfect CSI at the receiver will not be achieved in practice. Measurements of the channel path gains must be made at the receiver. Estimating the channel gains is typically accomplished by using preambles or by inserting pilot tones within a data frame, at the expense of a slight loss in rate. We note that for an OFDM system, it is not necessary to make channel measurements for every tone due to the correlation in the frequency domain. However, the quality of these measurements depends heavily on the placement of pilots (e.g., [39–41]).

Estimating the path gains can be accomplished in a very straightforward manner for orthogonal codes using pilot tones. We describe the procedure for the Alamouti Code.

We assume that two pilot tones x_1 and x_2 are transmitted, in the form of Equation 3.12, whose values are known to the receiver. The receiver obtains

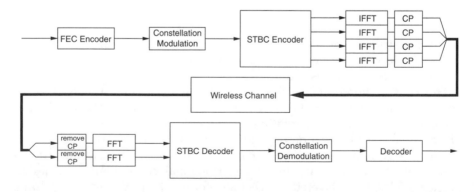

FIGURE 3.1
Implementation of a space–time block code in WiMAX.

the signals r_1 and r_2 where

$$\mathbf{r} = \begin{pmatrix} r_1 \\ r_2 \end{pmatrix} = \begin{pmatrix} x_1 & -x_2^* \\ x_2 & x_1^* \end{pmatrix} \begin{pmatrix} h_1 \\ h_2 \end{pmatrix} + \begin{pmatrix} z_1 \\ z_2 \end{pmatrix} = \mathbf{Xh} + \mathbf{z}$$

The optimal estimates for h_1, h_2 can be obtained by linear processing at the receiver, and are given by

$$\tilde{\mathbf{h}} = \begin{pmatrix} \tilde{h}_1 \\ \tilde{h}_2 \end{pmatrix} = \frac{1}{||\mathbf{x}||^2} \mathbf{X}^* \mathbf{r} = \begin{pmatrix} h_1 + \tilde{z}_1 \\ h_2 + \tilde{z}_2 \end{pmatrix}$$

where

$$\tilde{z}_1 = \frac{x_1 z_1 - x_2^* z_2}{||\mathbf{x}||^2}; \qquad \tilde{z}_2 = \frac{x_2 z_1 + x_1^* z_2}{||\mathbf{x}||^2}$$

These channel estimates can then be used to detect the next pair of code symbols. After the next code symbols are decoded, the channel estimate can be updated using those decoded symbols in place of the pilot symbols. When the channel variation is slow, the receiver improves stability of the decoding algorithm by averaging old and new channel estimates. This differential detection requires that the transmission begins with a pair of known symbols, and will perform within 3 dB of coherent detection where the CSI is perfectly known at the receiver.

3.4.3 A Differential Alamouti Code

In some circumstances, it is desirable to forgo the channel estimation module to keep the receiver complexity low. Channel estimates may also be unreliable due to motion. Under such circumstances, differential decoding algorithms become attractive despite their SNR loss from coherent decoding. In this section, we describe the differential encoding and decoding algorithm for the Alamouti Code [42,43]. We note that differential coding is not in the standard.

In Section 3.3.2, we described the Alamouti Code whose channel model was represented as

$$\begin{pmatrix} r_1 \\ -r_2^* \end{pmatrix} = \begin{pmatrix} h_1 & h_2 \\ -h_2^* & h_1^* \end{pmatrix} \begin{pmatrix} x_1 \\ -x_2^* \end{pmatrix} + \begin{pmatrix} z_1 \\ -z_2^* \end{pmatrix}$$

This expression can be manipulated further to

$$\begin{pmatrix} r_1 & r_2 \\ -r_2^* & r_1^* \end{pmatrix} = \begin{pmatrix} h_1 & h_2 \\ -h_2^* & h_1^* \end{pmatrix} \begin{pmatrix} x_1 & x_2 \\ -x_2^* & x_1^* \end{pmatrix} + \text{noise}$$

In the absence of noise, this can be rewritten as $\mathbf{R}(k) = \mathbf{HX}(k)$. Under quasi-static conditions, it follows that $\mathbf{R}^*(k-1)\mathbf{R}(k) = ||\mathbf{h}||^2 \mathbf{X}^*(k-1)\mathbf{X}(k)$.

Thus, we define the differential transmission rule

$$\mathbf{X}(k) = \overline{\mathbf{X}}^{-1}(k-1)\mathbf{U}(k) \tag{3.18}$$

where $\mathbf{U}(k)$ is the information matrix, which has Alamouti form. By setting the transmission rule in this manner, the receiver merely performs matrix multiplication to determine the transmitted matrix.

A method using differential Alamouti for ISI channels is also discussed in Ref. 43.

3.5   A Novel Quaternionic Space–Time Block Code

3.5.1   Code Construction

In the following section, we revisit the problem of designing orthogonal STBC for four transmit antennas. We present a novel full-rate full-diversity orthogonal STBC for four transmit antennas and in its applications in broadband wireless communication (WiMAX) environment. This code is constructed by means of a 2×2 array over the quaternions, thus resulting in a 4×4 array over the complex field $\mathbf{C}$. The code is orthogonal over $\mathbf{C}$ but is *not* linear. The structure of the code is a generalization of the 2×2 Alamouti Code [11], and reduces to it if the 2×2 quaternions in the code are replaced by complex numbers. For QPSK modulation, the code has no constellation expansion and enjoys a simple ML decoding algorithm. We also develop a differential encoding and decoding algorithm for this code. Another reason for our interest in orthogonal designs is that they limit the SNR loss incurred by differential decoding to its minimum of 3 dB from coherent decoding. A brief overview of equations is provided in the appendix.

We consider the STBC

$$\begin{bmatrix} p & q \\ -\bar{q} & \dfrac{\bar{q}\bar{p}q}{\|q\|^2} \end{bmatrix}$$

where the entries are quaternions. We may replace the quaternions p and q by the corresponding Alamouti 2×2 blocks to obtain a 4×4 STBC with complex entries.

$$\begin{bmatrix} \mathbf{P} & \mathbf{Q} \\ -\overline{\mathbf{Q}} & \dfrac{\overline{\mathbf{Q}}\,\overline{\mathbf{P}}\mathbf{Q}}{\|\mathbf{Q}\|^2} \end{bmatrix} \begin{bmatrix} \overline{\mathbf{P}} & -\mathbf{Q} \\ \overline{\mathbf{Q}} & \dfrac{\overline{\mathbf{Q}}\mathbf{P}\mathbf{Q}}{\|\mathbf{Q}\|^2} \end{bmatrix} = \left(\|p\|^2 + \|q\|^2 \right) \mathbf{I} \tag{3.19}$$

Observe that the rows of this code are orthogonal with respect to the standard inner product operation. Since QPSK signaling corresponds to choosing the quaternions p and q from the set $(\pm 1 \pm i \pm j \pm k)/2$, there is no constellation expansion because $(\bar{q}\bar{p}q)/\|q\|^2$ is always a quaternion of this same form. However, multiplication of quaternions is not commutative and it is not possible to have a 2×2 linear code over the quaternions with orthogonal rows and orthogonal columns [30].

3.5.2 Coherent Maximum Likelihood Decoding

We can represent the model in Equation 3.1 by quaternionic algebra. For simplicity, let us consider $M_r = 1$; all arguments can be easily generalized to $M_r > 1$. Consider the 2×2 complex matrices formed as

$$
\mathbf{R}_1 = \begin{bmatrix} \mathbf{r}(0) & \mathbf{r}(1) \\ -\bar{\mathbf{r}}(1) & \bar{\mathbf{r}}(0) \end{bmatrix}; \quad \mathbf{R}_2 = \begin{bmatrix} \mathbf{r}(2) & \mathbf{r}(3) \\ -\bar{\mathbf{r}}(3) & \bar{\mathbf{r}}(2) \end{bmatrix}
$$
$$
\mathbf{H}_1 = \begin{bmatrix} \mathbf{H}(1,1) & \mathbf{H}(1,2) \\ -\overline{\mathbf{H}}(1,2) & \overline{\mathbf{H}}(1,1) \end{bmatrix}; \quad \mathbf{H}_2 = \begin{bmatrix} \mathbf{H}(1,3) & \mathbf{H}(1,4) \\ -\overline{\mathbf{H}}(1,4) & \overline{\mathbf{H}}(1,3) \end{bmatrix}
\tag{3.20}
$$

where $\mathbf{H}(u, v)$ is the (u, v)th component of the channel matrix $\mathbf{H}$. Then we can rewrite Equation 3.1 for our code as

$$
[\mathbf{R}_1 \quad \mathbf{R}_2] = [\mathbf{H}_1 \quad \mathbf{H}_2] \begin{bmatrix} \mathbf{P} & \mathbf{Q} \\ -\overline{\mathbf{Q}} & \dfrac{\overline{\mathbf{Q}}\,\overline{\mathbf{P}}\mathbf{Q}}{\|\mathbf{Q}\|^2} \end{bmatrix} + [\mathbf{Z}_1 \quad \mathbf{Z}_2]
\tag{3.21}
$$

where the noise vectors are also replaced by corresponding quaternionic matrices of the forms given in Equation 3.20.

From Equation 3.21, the ML decoding rule is given by*

$$
\left\{ \hat{\mathbf{P}}, \hat{\mathbf{Q}} \right\} = \arg \min_{\mathbf{P},\mathbf{Q}} \left\| [\mathbf{R}_1 \quad \mathbf{R}_2] - [\mathbf{H}_1 \quad \mathbf{H}_2] \begin{bmatrix} \mathbf{P} & \mathbf{Q} \\ -\overline{\mathbf{Q}} & \dfrac{\overline{\mathbf{Q}}\,\overline{\mathbf{P}}\mathbf{Q}}{\|\mathbf{Q}\|^2} \end{bmatrix} \right\|^2
$$
$$
= \arg \max_{\mathbf{P},\mathbf{Q}} \Re \left\{ \operatorname{trace} \left([\mathbf{R}_1 \quad \mathbf{R}_2] \begin{bmatrix} \overline{\mathbf{P}} & -\mathbf{Q} \\ \overline{\mathbf{Q}} & \dfrac{\overline{\mathbf{Q}}\mathbf{P}\mathbf{Q}}{\|\mathbf{Q}\|^2} \end{bmatrix} \begin{bmatrix} \overline{\mathbf{H}}_1 \\ \overline{\mathbf{H}}_2 \end{bmatrix} \right) \right\}
\tag{3.22}
$$

---

* Assuming that $\|\mathbf{P}\|$ and $\|\mathbf{Q}\|$ are constant.

3.5.3   An Efficient Decoder

We can write Equation 3.21 in quaternionic algebra as follows

$$[r_1 \quad r_2] = [h_1 \quad h_2] \begin{bmatrix} p & q \\ -\bar{q} & \dfrac{\bar{q}\bar{p}q}{\|q\|^2} \end{bmatrix} + [z_1 \quad z_2] \tag{3.23}$$

where we have defined h_1, h_2 as the quaternions corresponding to the matrices $\mathbf{H}_1, \mathbf{H}_2$ given in Equation 3.21. Being inspired by the simplicity of decoding scheme of the standard Alamouti Code [11] through linear combinations of the received signals, we generalize the idea and derive the following four expressions by linearly combining the received signals in Equation 3.23. Interested readers can find the detailed derivations of the linear combination process in Ref. 44.

$$\Re(\tilde{r}_1) = \Re\left(h_1\bar{r}_1 + r_2\bar{h}_2\right) = \left(\|h_1\|^2 + \|h_2\|^2\right)p_0 + \Re(\tilde{z}_0) \tag{3.24}$$

$$\Re(\tilde{r}_2) \stackrel{\text{def}}{=} \Re\left[h_1\bar{i}\bar{r}_1 + r_2(i\mathbf{T}_q)\bar{h}_2\right] = \left(\|h_1\|^2 + \|h_2\|^2\right)p_1 + \Re(\tilde{z}_1) \tag{3.25}$$

$$\Re(\tilde{r}_3) \stackrel{\text{def}}{=} \Re\left[h_1\bar{j}\bar{r}_1 + r_2(j\mathbf{T}_q)\bar{h}_2\right] = \left(\|h_1\|^2 + \|h_2\|^2\right)p_2 + \Re(\tilde{z}_2) \tag{3.26}$$

and

$$\Re(\tilde{r}_4) \stackrel{\text{def}}{=} \Re\left[h_1k\bar{r}_1 + r_2(k\mathbf{T}_q)\bar{h}_2\right] = \left(\|h_1\|^2 + \|h_2\|^2\right)p_3 + \Re(\tilde{z}_3) \tag{3.27}$$

Decoding proceeds as follows. First, p_0 is calculated by applying a hard slicer to the left-hand side of Equation 3.24. Next, as discussed in the appendix, there are eight choices for the transformation $\mathbf{T}_q$ where each can be used to calculate a candidate for the triplet (p_1, p_2, p_3) by applying a hard slicer to the left-hand sides of Equations 3.25 through 3.27. For each choice of $\mathbf{T}_q$, there are two choices of q (sign ambiguity). Finally, the 16 candidates for (p, q) are compared using the ML metric in Equation 3.22 to obtain the decoded QPSK information symbols. We have proved that the statistics $\Re(\tilde{r}_1)$ through $\Re(\tilde{r}_4)$ are sufficient for ML decoding [44]. In addition, we emphasize that there is no loss of optimality in applying the hard QPSK slicer operation to Equation 3.24 through 3.27 since the noise samples are zero-mean uncorrelated Gaussian.

3.5.4   A Differential Quaternionic Code

Our starting point is the input–output relationship in Equation 3.23 that can be written in compact matrix notation as follows

$$\mathbf{r}^{(k)} = \mathbf{h}\mathbf{C}^{(k)} + \mathbf{z}^{(k)} \tag{3.28}$$

Consider the following differential encoding rule

$$\mathbf{C}^{(k)} = \mathbf{C}^{(k-1)}\mathbf{U}^{(k)} \tag{3.29}$$

where the information matrix

$$\mathbf{U}^{(k)} = \begin{bmatrix} \mathbf{P} & \mathbf{Q} \\ -\overline{\mathbf{Q}} & \dfrac{\overline{\mathbf{Q}}\,\mathbf{P}\mathbf{Q}}{\|\mathbf{Q}\|^2} \end{bmatrix}$$

Therefore, we have

$$\mathbf{r}^{(k)} = \mathbf{h}\mathbf{C}^{(k-1)}\mathbf{U}^{(k)} + \mathbf{z}^{(k)} \tag{3.30}$$

from which we can write

$$\begin{aligned} \mathbf{r}^{(k)} &= \left(\mathbf{r}^{(k-1)} - \mathbf{z}^{(k-1)}\right)\mathbf{U}^{(k)} + \mathbf{z}^{(k)} \\ &= \mathbf{r}^{(k-1)}\mathbf{U}^{(k)} + \underbrace{\mathbf{z}^{(k)} - \mathbf{z}^{(k-1)}\mathbf{U}^{(k)}}_{\tilde{\mathbf{z}}^{(k)}} \end{aligned} \tag{3.31}$$

This equation has identical form to the received signal equation in the coherent case **except** for the two main differences.

- The previous output vector $\mathbf{r}^{(k-1)}$ in Equation 3.31 plays the role of the channel coefficient vector and is known at the receiver.
- Since $\mathbf{U}^{(k)}$ is a unitary matrix by construction, the equivalent noise vector $\tilde{\mathbf{z}}^{(k)}$ will also be zero-mean white Gaussian (such as $\mathbf{z}^{(k)}$ and $\mathbf{z}^{(k-1)}$) but with twice the variance.

Hence, the same efficient ML coherent decoding algorithm applies in the differential case as well but at an additional 3 dB performance penalty at high SNR.

3.6 Simulation Results

We present simulation results on the performance of our proposed STBC with the efficient ML decoding algorithm. We assume QPSK modulation, a single antenna at the receiver (unless otherwise stated), and no CSI at the transmitter.

We start by investigating the resulting performance degradation when the assumption of perfect CSI at the receiver is not satisfied. We consider two scenarios. In the first scenario, no CSI is available at the receiver and the differential encoding/decoding scheme of Section 3.5.4 is used. Figure 3.2 shows that the SNR penalty from coherent decoding (with perfect CSI) is 3 dB at high SNR. In the second scenario, the coherent ML decoder uses estimated CSI acquired by transmitting a pilot codeword of the same quaternionic structure and using a simple matched filter operation at the receiver to calculate the CSI vector. Figure 3.3 shows that the performance loss due to channel estimation is about 2–3 dB, which is comparable to the differential technique.

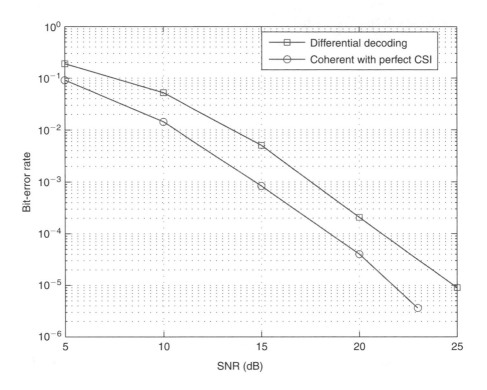

FIGURE 3.2
Performance comparison between coherent and differential decoding in quasi-static fading.

Next, we compare the performance of both the schemes in a time-varying channel. The pilot-based channel estimation scheme will suffer performance degradation since the channel estimate will be outdated due to the Doppler effect. To mitigate this effect, we need to increase the frequency of pilot code-word insertion as the Doppler frequency increases, which in turn increases the training overhead. We assume a fixed pilot insertion rate of one every 20 codewords; that is, a training overhead of only 5%. Similarly, the differential scheme will also suffer performance degradation since the assumption of a constant channel over two consecutive codewords (i.e., eight symbol intervals) is no longer valid. Figure 3.4 shows that for high mobile speeds (≥ 60 mph), an error floor occurs for both schemes.

Both schemes achieve comparable performance for low (pedestrian, ≤ 5 mph) speeds, but the pilot-based scheme performs better at moderate to high speeds at the expense of a more complex receiver (to perform channel estimation) and the pilot transmission overhead.

To investigate the performance of our proposed quaternionic code in a wireless broadband environment, we assume the widely-used Stanford University Interim (SUI) channel models [45] where each of the three-tap SUI channel models is defined for a particular terrain type with varying degree of Ricean

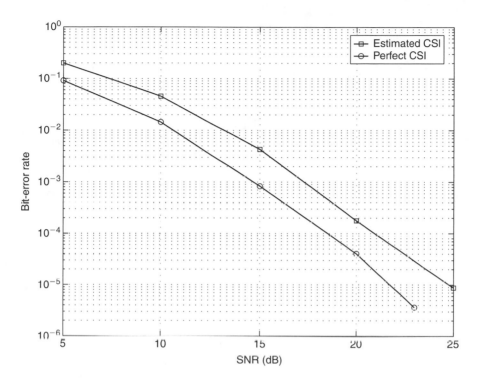

FIGURE 3.3
Performance comparison between perfect CSI and estimated CSI in quasi-static fading.

fading **K** factors and Doppler frequency. We combine our quaternionic STBC with OFDM transmission where each codeword is now transmitted over four consecutive OFDM symbol durations (for each tone). In our simulations, we use 256 subcarriers and a cyclic prefix length of 64 samples. We simultaneously transmit 256 codewords from four transmit antennas over four OFDM symbols and assume that the channel remains fixed over that period. We also use a Reed–Solomon RS(255, 163) outer code and frequency-interleave of the coded data before transmitting through the channel. This simulation model is not compliant with WiMAX, but it captures the concept of the technologies used in WiMAX.

Figure 3.5 illustrates the significant performance gains achieved by our proposed 4-TX STBC in the 802.16 environment as compared to SISO transmission. To put these SNR gains in perspective, at BER $= 10^{-3}$, these SNR gains translate to a 50% increase in the cell coverage area assuming one receive Antenna. By adding a second receive antenna, the percentage increase becomes 166%.*

---

* These calculations assume a path loss exponent of 4, which is recommended for the SUI-3 channel model with a Base Station height of 50 m [45].

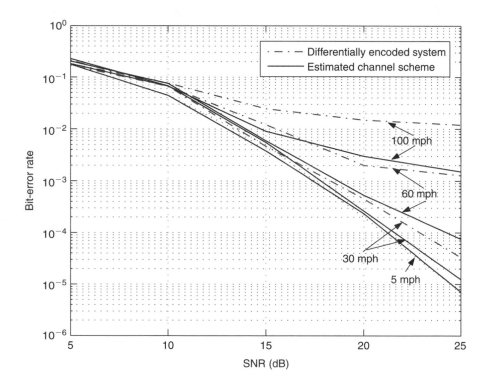

FIGURE 3.4
Performance comparison between differential and pilot-based decoding schemes in time-varying channel.

Appendix: Quaternions

Quaternions are a noncommutative extension of the complex numbers. In the mid-nineteenth century, Hamilton discovered quaternions and was so pleased that he immediately carved the following message into the Brougham Bridge in Dublin [46].

$$i^2 = j^2 = k^2 = ijk = -1 \tag{3.32}$$

This equation is the fundamental equation for the quaternions. A quaternion can be written uniquely as a linear combination of the four basis quaternions $1, i, j, k$, over the reals:

$$q \overset{\text{def}}{=} q_0 + q_1 i + q_2 j + q_3 k \tag{3.33}$$

It is a noncommutative group, as can be seen by the relations $ij = -ji = k$, $jk = -kj = i$, and $ki = -ik = j$. These relations follow from Equation 3.32 and associativity.

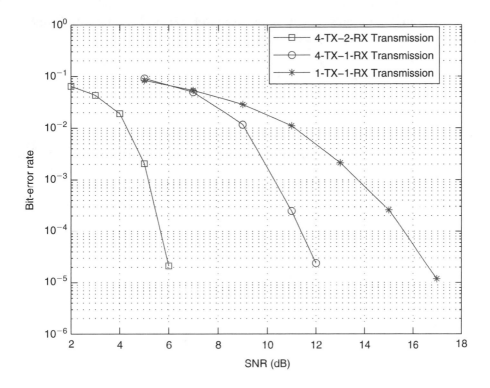

FIGURE 3.5
Performance comparison between our proposed code (with one and two receive antennas) and SISO transmission. Both are combined with OFDM in an 802.16 scenario.

Quaternions can be viewed as a 4×4 matrix algebra over the real numbers $\mathbb{R}$, where right multiplication by the quaternion q is described by

$$x_0 + x_1 i + x_2 j + x_3 k \equiv [x_0 \ x_1 \ x_2 \ x_3] \rightarrow [x_0 \ x_1 \ x_2 \ x_3] \begin{bmatrix} q_0 & q_1 & q_2 & q_3 \\ -q_1 & q_0 & -q_3 & q_2 \\ -q_2 & q_3 & q_0 & -q_1 \\ -q_3 & -q_2 & q_1 & q_0 \end{bmatrix}$$

(3.34)

The conjugate quaternion $\bar{q}$ is given by $\bar{q} \stackrel{\text{def}}{=} q_0 - q_1 i - q_2 j - q_3 k$ and we have

$$q\bar{q} = \|q\|^2 = q_0^2 + q_1^2 + q_2^2 + q_3^2$$

(3.35)

We may also view quaternions as pairs of complex numbers, where the product of quaternions (v, w) and (v', w') is given by

$$(v, w)(v', w') = (vv' - \bar{w}'w, vw' + \bar{v}'w)$$

(3.36)

These are Hamilton's biquaternions (see Ref. 47), and right multiplication by the biquaternion (v, w) is described by

$$x_0 + x_1 i + y_0 j + y_1 k \equiv [x \quad y] \rightarrow [x \quad y] \begin{bmatrix} v & w \\ -\bar{w} & \bar{v} \end{bmatrix} \tag{3.37}$$

The matrices

$$\begin{bmatrix} 1 & 0 \\ 0 & 1 \end{bmatrix}; \quad \begin{bmatrix} i & 0 \\ 0 & -i \end{bmatrix}; \quad \begin{bmatrix} 0 & 1 \\ -1 & 0 \end{bmatrix}; \quad \begin{bmatrix} 0 & i \\ i & 0 \end{bmatrix} \tag{3.38}$$

describe right multiplication by 1, i, j, and k, respectively.

There is an isomorphism between the quaternions q, 4×4 real matrices, and 2×2 complex matrices:

$$q \cong \begin{bmatrix} q_0 & q_1 & q_2 & q_3 \\ -q_1 & q_0 & -q_3 & q_2 \\ -q_2 & q_3 & q_0 & -q_1 \\ -q_3 & -q_2 & q_1 & q_0 \end{bmatrix} \cong \begin{bmatrix} q^c(0) & q^c(1) \\ -\bar{q}^c(1) & \bar{q}^c(0) \end{bmatrix} = \mathbf{Q} \tag{3.39}$$

where $q^c(0)$, $q^c(1) \in \mathbf{C}$, and $q^c(0) = q_0 + iq_1$, $q^c(1) = q_2 + iq_3$. Therefore, we can interchangeably use the matrix representation for the quaternions to demonstrate their properties. We will represent the 2×2 complex version of q by $\mathbf{Q}$. The norms have the following relationship:

$$||q||^2 = q_0^2 + q_1^2 + q_2^2 + q_3^2 = |q^c(0)|^2 + |q^c(1)|^2 = ||\mathbf{Q}||^2 \tag{3.40}$$

The matrix representing right multiplication by the biquaternion (v, w) is the 2×2 STBC introduced by Alamouti [11]. Note that the rows and columns are orthogonal with respect to the standard inner product

$$[x \quad y] \cdot [x' \quad y'] = x\bar{x}' + y\bar{y}' \tag{3.41}$$

There is a classical correspondence between unit quaternions and rotations in $\mathbf{R}^3$ given by

$$q \longrightarrow \mathbf{T}_q : p \longrightarrow \bar{q}pq \tag{3.42}$$

where we have identified vectors in $\mathbf{R}^4$ with quaternions $p = p_0 + p_1 i + p_2 j + p_3 k$ [46]. The transformation $\mathbf{T}_q$ fixes the real part $\Re(p)$ of the quaternion p, and if $q = q_0 + q_1 i + q_2 j + q_3 k$, then $\mathbf{T}_q$ describes rotation about the axis (q_1, q_2, q_3) through an angle 2θ where $\cos(\theta) = q_0$ and $\sin(\theta) = \sqrt{q_1^2 + q_2^2 + q_3^2}$. For example, if $q = \pm(1 + i + j + k)/2$, the transformation $\mathbf{T}_q$ and its effect on i, j, and k are as follows:

$$\mathbf{T}_q = \begin{bmatrix} 1 & 0 & 0 & 0 \\ 0 & 0 & 0 & 1 \\ 0 & 1 & 0 & 0 \\ 0 & 0 & 1 & 0 \end{bmatrix}; \quad \begin{array}{l} i \rightarrow k \\ j \rightarrow i \\ k \rightarrow j \end{array} \tag{3.43}$$

The eight transformations $\mathbf{T}_q$ together with their effect on i, j, and k can be found in Ref. 44. A transformation $\mathbf{T}_q$ *either* maps $i \to \pm k$; $k \to \pm j$; $j \to \pm i$ *or* maps $i \to \pm j$; $j \to \pm k$; $k \to \pm i$. In both cases, the product of the signs is equal to 1.

References

1. G. J. Foschini, Layered space-time architecture for wireless communication in a fading environment when using multi-element antennas, *Bell Labs Tech. J.*, vol. 1, no. 2, pp. 41–59, Fall 1996.
2. E. Telatar, Capacity of multi-antenna Gaussian channels, *European Trans. Telecommun.*, vol. 10, no. 6, pp. 585–595, Nov./Dec. 1999.
3. Y. Li and G. L. Stüber, *Orthogonal Frequency Division Multiplexing for Wireless Communications*, New York: Springer, 2006.
4. *IEEE Standard for Local and Metropolitan Area Networks Part 16: Air Interface for Fixed Broadband Wireless Access Systems*, IEEE Std. 802.16-2004, 2004.
5. *IEEE Std 802.16e-2005 and IEEE Std 802.16-2004/Cor 1-2005. Amendment and Corrigendum to IEEE Std 802.16-2004*, IEEE, Std. 802.16e-2005, 2005.
6. V. Tarokh, N. Seshadri, and A. R. Calderbank, Space-time codes for high data rate wireless communication: Performance criterion and code construction, *IEEE Trans. Inform. Theory*, vol. 44, no. 2, pp. 744–765, Mar. 1998.
7. S. Parkvall, M. Karlsson, M. Samuelsson, L. Hedlund, and B. Göransson, Transmit diversity in WCDMA: Link and system level results, in *Veh. Techn. Conf. Proc., VTC 2000-Spring*, Tokyo, Japan, May 2000.
8. A. F. Naguib, V. Tarokh, N. Seshadri, and A. R. Calderbank, A space-time coding modem for high-data-rate wireless communications, *IEEE J. Select. Areas Commun.*, vol. 16, no. 8, pp. 1459–1478. Oct. 1998.
9. V. Tarokh, A. Naguib, N. Seshadri, and A. R. Calderbank, Space-time codes for high data rate wireless communication: Performance criteria in the presence of channel estimation errors, mobility, and multiple paths, *IEEE Trans. Commun.*, vol. 47, no. 2, pp. 199–207, Feb. 1999.
10. WiMAX forum [Online]. Available: http//www.wimaxforum.org, 2006.
11. S. M. Alamouti, A simple transmit diversity technique for wireless communications, *IEEE J. Select. Areas Commun.*, vol. 16, no. 8, pp. 1451–1458, Oct. 1998.
12. J. Guey, M. P. Fitz, M. R. Bell, and W. Kuo, Signal design for transmitter diversity wireless communication systems over Rayleigh fading channels, *IEEE Trans. Commun.*, vol. 47, no. 4, pp. 527–537, Apr. 1999.
13. H. Lu, Y. Wang, P. V. Kumar, and K. M. Chugg, Remarks on space-time codes including a new lower bound and an improved code, *IEEE Trans. Inform. Theory*, vol. 49, no. 10, pp. 2752–2757, Oct. 2003.
14. H. Jafarkhani, *Space–Time Coding: Theory and Practice*. Cambridge University Press, New York, 2005.
15. J. Chui and A. R. Calderbank, Effective coding gain for space–time codes, in *Proc. IEEE Int. Symp, Info. Theory (ISIT)*, Seattle, WA, July 2006.

16. L. Zheng and D. N. Tse, Diversity and multiplexing: A fundamental trade-off in multiple-antenna channels, *IEEE Trans. Inform. Theory*, vol. 49, no. 5, pp. 1073–1096, May 2003.

17. S. N. Diggavi, N. Al-Dhahir, A. Stamoulis, and A. R. Calderbank, Great expectations: The value of spatial diversity to wireless networks, in *Proc. IEEE*, vol. 92, no. 2, pp. 219–270, Feb. 2004, invited paper.

18. L. Grokop and D. Tse, Diversity/multiplexing tradeoff in ISI channels, in *Proc. IEEE Int. Symp. Info. Theory (ISIT)*, June/July 2004.

19. G. J. Foschini, G. Golden, R. Valenzuela, and P. Wolniansky, Simplified processing for high spectral efficiency wireless communication employing multi-element arrays, *IEEE J. Select Areas Commun.*, vol. 17, no. 11, pp. 1841–1852, Nov. 1999.

20. W. Choi, R. Negi, and J. M. Cioffi, Combined ML and DFE decoding for the V-BLAST system, in *Proc. ICC*, vol. 3, pp. 1243–1248, June 2003.

21. A. F. Naguib and N. Seshadri, Combined interference cancellation and ML decoding of space–time block codes, in *Proc. Comm. Theory Mini-conference, Held in Conjunction with Globecomm '98*, Sydney, Australia, 1998, pp. 7–15.

22. S. Sirianunpiboon, A. R. Calderbank, and S. D. Howard, Space-polarization-time codes for diversity gains across line of sight and rich scattering environments, *Trans. on Info. Theory*, 2006 (submitted).

23. H. Yao and G. W. Wornell, Achieving the full mimo diversity-multiplexing frontier with rotation-based space-time codes, in *Proc. Allerton Conf. Commun., Control, Comput.*, Oct. 2003.

24. P. Dayal and M. K. Varanasi, An optimal two transmit antenna space-time code and its stacked extensions, *IEEE Trans. Inform. Theory*, vol. 51, no. 12, pp. 4348–4355, Dec. 2005.

25. J.-C. Belfiore, G. Rekaya, and E. Viterbo, The golden code: A 2×2 full-rate space-time code with nonvanishing determinants, *IEEE Trans. Inform. Theory*, vol. 51, no. 4, pp. 1432–1436, Apr. 2005.

26. E. Viterbo and J. Boutros, A universal lattice code decoder for fading channels, *IEEE Trans. Inform. Theory*, vol. 45, no. 5, pp. 1639–1642, July 1999.

27. M. O. Damen, A. Chkeif, and J.-C. Belfiore, Lattice code decoder for space-time codes, *IEEE Commun. Lett.*, vol. 4, no. 5, pp. 161–163, May 2000.

28. U. Fincke and M. Pohst, Improved methods for calculating vectors of short length in a lattice, including a complexity analysis, *Math. Comp.*, vol. 44, no. 170, pp. 463–471, Apr. 1985.

29. S. J. Lee et al., A space-time code with full-diversity and rate 2 for 2 transmit antenna transmission, in *IEEE 802.16 Session #34*, 2004, Contribution IEEE C802.16e-04/434r2, Nov. 2004.

30. V. Tarokh, H. Jafarkhani, and A. R. Calderbank, Space-time block codes from orthogonal designs, *IEEE Trans. Inform. Theory*, vol. 45, no. 5, pp. 1456–1467, July 1999.

31. H. Wang and X.-G. Xia, Upper bounds of rates of complex orthogonal space-time block codes, *IEEE Trans. Inform. Theory*, vol. 49, no. 10, pp. 2788–2796, Oct. 2003.

32. H. Jafarkhani, A quasi-orthogonal space-time block code, *IEEE Trans. Commun.*, vol. 49, no. 1, pp. 1–4, Jan. 2001.

33. N. Sharma and C. B. Papadias, Improved quasi-orthogonal codes through constellation rotation, *IEEE Trans. Commun.*, vol. 51, no. 3, pp. 332–335, Mar. 2003.

34. H. Jafarkhani and N. Seshadri, Super-orthogonal space-time trellis codes, *IEEE Trans. Inform. Theory*, vol. 49, no. 4, pp. 937–950, Apr. 2003.

35. B. Hassibi and B. Hochwald, High-rate codes that are linear in space and time, *IEEE Trans. Inform. Theory*, vol. 48, no. 7, pp. 1804–1824, July 2002.

36. H. E. Gamal and A. R. Hammons, Jr., A new approach to layered space-time coding and signal processing, *IEEE Trans. Inform. Theory*, vol. 47, no. 6, pp. 2321–2334, Sept. 2001.

37. M. O. Damen, H. El Gamal, and N. C. Beaulieu, Linear threaded algebraic space-time constellations, *IEEE Trans. Inform. Theory*, vol. 49, no. 10, pp. 2372–2388, Oct. 2003.

38. F. Oggier, G. Rekaya, J.-C. Belfiore, and E. Viterbo, Perfect space-time block codes, *IEEE Trans. Inform. Theory*, vol. 52, no. 9, pp. 3885–3902, Sept. 2006.

39. J. Cavers, An analysis of pilot symbol assisted modulation for Rayleigh fading channels (mobile radio), *IEEE Trans. Veh. Technol.*, vol. 40, no. 4, pp. 686–693, Nov. 1991.

40. R. Negi and J. Cioffi, Pilot tone selection for channel estimation in a mobile OFDM system, *IEEE Trans. Consumer Electron.*, vol. 44, no. 3, pp. 1122–1128, Aug. 1998.

41. S. Ohno and G. B. Giannakis, Capacity maximizing MMSE-optimal pilots for wireless OFDM over frequency-selective block Rayleigh-fading channels, *IEEE Trans. Inform. Theory*, vol. 50, no. 9, pp. 2138–2145, Sept. 2004.

42. V. Tarokh and H. Jafarkhani, A differential detection scheme for transmit diversity, *IEEE J. Select. Areas Commun.*, vol. 18, no. 7, pp. 1169–1174, July 2000.

43. S. N. Diggavi, N. Al-Dhahir, A. Stamoulis, and A. R. Calderbank, Differential space-time coding for frequency-selective channels, *IEEE Commun. Lett.*, vol. 6, no. 6, pp. 253–255, June 2002.

44. R. Calderbank, S. Das, N. Al-Dhahir, and S. Diggavi, Construction and analysis of a new quaternionic space-time code for 4 transmit antennas, *Commun. Inform. Syst.* (Special Issue Dedicated to the 70th Birthday of Thomas Kailath: Part I), vol. 5, no. 1, pp. 97–122, 2005.

45. V. Erceg, K. V. S. Hari, M. S. Smith, D. S. Baum, et al., Channel models for fixed wireless applications, in *IEEE 802.16 Broadband Wireless Access Working Group*, *IEEE 802.16a-03/01*, 2003.

46. J. H. Conway and D. Smith, *On Quaternions and Octonions*. AK Peters, Ltd., Wellesley, Massachusetts, 2003.

47. B. L. van der Waerden, *A History of Algebra: From Al-Khwarizmi to Emmy Noether*. New York: Springer, 1985.

4

Exploiting Diversity in MIMO-OFDM Systems for Broadband Wireless Communications

Weifeng Su, Zoltan Safar, and K. J. Ray Liu

CONTENTS

4.1 Introduction

WiMAX is a broadband wireless solution that is likely to play an important role in providing ubiquitous voice and data services to millions of users in the near future, both in rural and urban environment. It is based on the IEEE 802.16 air interface standard suite [1,2], which provides the wireless technology for both fixed, nomadic, and mobile data access. Wireless systems have the capacity to cover large geographic areas without the need for costly cable infrastructure to each service access point, so WiMAX has the potential to prove to be a cost-effective and quickly deployable alternative to cabled networks, such as fiber optic links, cable modems, or digital subscriber lines (DSL).

The driving force behind the development of the WiMAX system has been the desire to satisfy the emerging need for high data rate applications, e.g., voice over IP, video conferencing, interactive gaming, and multimedia streaming. However, recently performed system-level simulation results indicate that the performance of the current WiMAX system may not be able to satisfy such needs. In Ref. 3, the downlink (DL) performance of a 10 MHz WiMAX system was evaluated in a 1/1 frequency reuse scenario with two transit and two receive antennas, and 10–14 Mbit/s average total sector throughput was obtained. The average DL cell throughput of a 5 MHz WiMAX system in Ref. 4 was found to be around 5 Mbit/s, also with two transit and two receive antennas and 1/1 frequency reuse. As a consequence, it seems that further performance improvement is necessary to be able to support interactive multimedia applications that require the user data rates in excess of 0.5–1 Mbit/s. The authors of Ref. 4 provide a list of techniques to achieve this: hybrid automatic repeat-request (ARQ), interference cancellation, adaptive per-subcarrier power allocation, and frequency-domain scheduling.

This list can be appended with one more item: improved multiantenna coding techniques. So far, the only open-loop multiple-input multiple-output (MIMO) coding method adopted by the WiMAX forum has been the Alamouti's 2×2 orthogonal design [7] (in different variants). Since for high-mobility users the closed-loop transmission techniques, such as frequency-domain scheduling or beam forming, are not available due to the feedback loop delay, more powerful open-loop MIMO coding methods could help to achieve even higher data rates than it is possible now with the current WiMAX system. One way to characterize the performance of MIMO systems is by their *diversity* order, which is the asymptotic slope of the bit-error rate (BER) versus signal-to-noise ratio (SNR) curve, i.e., it describes how fast the BER decreases as the SNR increases. The achieved diversity order depends on both the MIMO channel (its spatial, spectral, and temporal structure) and on the applied MIMO coding and decoding method, which can exploit a part or all of the available diversity in the MIMO channel.

The problem of designing MIMO coding and modulation techniques to improve the performance of wireless communication systems has attracted considerable attention in both industry and academia. In case of narrowband wireless communications, where the fading channel is frequency nonselective, abundant coding methods [5–10], termed as space–time (ST) codes, have been proposed to exploit the spatial and temporal diversities. In case of broadband wireless communications, where the fading channel is frequency-selective, orthogonal frequency division multiplexing (OFDM) modulation can be used to convert the frequency-selective channel into a set of parallel flat-fading channels, providing high spectral efficiency and eliminating the need for high-complexity equalization algorithms. To have the "best of both worlds," MIMO systems and OFDM modulation can be combined, resulting in MIMO-OFDM systems. This combination seemed so attractive that several communication systems, including the MIMO option in WiMAX, were based on it.

There are two major coding approaches for MIMO-OFDM systems. One is the space–frequency (SF) coding approach, where coding is applied within each OFDM block to exploit the spatial and frequency diversities. The other one is the space–time–frequency (STF) coding approach, where the coding is applied across multiple OFDM blocks to exploit the spatial, temporal, and frequency diversities. Early works on SF coding [11–16] used ST codes directly as SF codes, i.e., previously existing ST codes were used by replacing the time domain with the frequency domain (OFDM tones). The performance criteria for SF-coded MIMO-OFDM systems were derived in Refs. 16 and 17, and the maximum achievable diversity was found to be LM_tM_r, where M_t and M_r are the number of transmit and receive antennas, respectively, and L is the number of delay paths in frequency-selective fading channels. It has been shown in Ref. 17 that the way of using ST codes directly as SF codes can achieve only the spatial diversity, but not the full spatial and frequency diversity LM_tM_r. Later, in Refs. 18–20, systematic SF code design methods were proposed that could guarantee to achieve the maximum diversity.

One may also consider STF coding across multiple OFDM blocks to exploit all of the spatial, temporal, and frequency diversities. The STF coding strategy was first proposed in Ref. 21 for two transmit antennas and further developed in Refs. 22–24 for multiple transmit antennas. Both Refs. 21 and 24 assumed that the MIMO channel stays constant over multiple OFDM blocks; however, STF coding under this assumption cannot provide any additional diversity compared to the SF coding approach [26]. In Ref. 23, an intuitive explanation on the equivalence between antennas and OFDM tones was presented from the viewpoint of channel capacity. In Ref. 22, the performance criteria for STF codes were derived, and an upper bound on the maximum achievable diversity order was established. However, there was no discussion in Ref. 22 whether the upper bound can be achieved or not, and the proposed STF codes were not guaranteed to achieve the full spatial, temporal, and frequency diversities. Later, in Ref. 26, we proposed a systematic method to design full-diversity STF codes for MIMO-OFDM systems.

In this chapter, we review SF/STF code design criteria and summarize our findings on SF/STF coding for MIMO-OFDM systems [19,20,26]. The chapter is organized as follows. First, we describe a general MIMO-OFDM system model and review code design criteria. Second, we introduce two SF code design methods that can guarantee to achieve full spatial and frequency diversity based on Refs. 19 and 20. Then, we summarize our results on STF coding based on Ref. 26. Finally, some simulation results are presented and some conclusions are drawn.

4.2 MIMO-OFDM System Model and Code Design Criteria

We first describe a general STF-coded MIMO-OFDM system and discuss its performance criteria. Since SF coding, where coding is applied within each OFDM block, is a special case of STF coding, the performance criteria of SF-coded MIMO-OFDM systems can be easily obtained from that of STF codes, as shown at the end of this section.

4.2.1 System Model

We consider a general STF-coded MIMO-OFDM system with M_t transmit antennas, M_r receive antennas, and N subcarriers. Suppose that the frequency-selective fading channels between each pair of transceiver antennas have L independent delay paths and the same power delay profile. The MIMO channel is assumed to be constant over each OFDM block, but it may vary from one OFDM block to another. At the kth OFDM block, the channel coefficient from transmit antenna i to receive antenna j at time τ can be modeled as

$$h_{i,j}^k(\tau) = \sum_{l=0}^{L-1} \alpha_{i,j}^k(l)\delta(\tau - \tau_l) \tag{4.1}$$

where τ_l is the delay and $\alpha_{i,j}^k(l)$ the complex amplitude of the lth path between transmit antenna i and receive antenna j. The $\alpha_{i,j}^k(l)$ are modeled as zero-mean, complex Gaussian random variables with variances δ_l^2 and $\sum_{l=0}^{L-1} \delta_l^2 = 1$. From Equation 4.1, the frequency response of the channel is given by $H_{i,j}^k(f) = \sum_{l=0}^{L-1} \alpha_{i,j}^k(l)e^{-j2\pi f \tau_l}$ and $j = \sqrt{-1}$.

We consider STF coding across M_t transmit antennas, N OFDM subcarriers, and K consecutive OFDM blocks (the $K = 1$ case corresponds to SF coding). Each STF codeword can be expressed as a $KN \times M_t$ matrix

$$C = \begin{bmatrix} C_1^T & C_2^T & \cdots & C_K^T \end{bmatrix}^T \tag{4.2}$$

where the channel symbol matrix C_k is given by

$$C_k = \begin{bmatrix} c_1^k(0) & c_2^k(0) & \cdots & c_{M_t}^k(0) \\ c_1^k(1) & c_2^k(1) & \cdots & c_{M_t}^k(1) \\ \vdots & \vdots & \ddots & \vdots \\ c_1^k(N-1) & c_2^k(N-1) & \cdots & c_{M_t}^k(N-1) \end{bmatrix} \tag{4.3}$$

in which $c_i^k(n)$ is the channel symbol transmitted over the nth subcarrier by transmit antenna i in the kth OFDM block. The STF code is assumed to satisfy the energy constraint $E||C||_F^2 = KNM_t$, where $||C||_F$ is the Frobenius norm of C. During the kth OFDM block period, the transmitter applies an N-point IFFT to each column of the matrix C_k. After appending a cyclic prefix, the OFDM symbol corresponding to the ith ($i = 1, 2, \ldots, M_t$) column of C_k is transmitted by transmit antenna i.

At the receiver, after removing the cyclic prefix and applying FFT, the received signal at the nth subcarrier at receive antenna j in the kth OFDM block is given by

$$y_j^k(n) = \sqrt{\frac{\rho}{M_t}} \sum_{i=1}^{M_t} c_i^k(n) H_{i,j}^k(n) + z_j^k(n) \tag{4.4}$$

where

$$H_{i,j}^k(n) = \sum_{l=0}^{L-1} \alpha_{i,j}^k(l) e^{-j2\pi n \Delta f \tau_l} \tag{4.5}$$

is the channel frequency response at the nth subcarrier between transmit antenna i and receive antenna j; $\Delta f = 1/T$ the subcarrier separation in the frequency domain; and T the OFDM symbol period. We assume that the channel state information $H_{i,j}^k(n)$ is known at the receiver, but not at the transmitter. In Equation 4.4, $z_j^k(n)$ denotes the additive white complex Gaussian noise with zero-mean and unit variance at the nth subcarrier at receive antenna j in the kth OFDM block. The factor $\sqrt{\rho/M_t}$ in Equation 4.4 ensures that ρ is the average SNR at each receive antenna.

4.2.2 Code Design Criteria

We discuss the STF code design criteria based on the pairwise error probability of the system. The channel frequency response vector between transmit antenna i and receive antenna j over K OFDM blocks will be denoted by

$$H_{i,j} = [H_{i,j}(0) \ H_{i,j}(1) \ \cdots \ H_{i,j}(KN - 1)]^T \tag{4.6}$$

where we use the notation $H_{i,j}((k-1)N + n) \triangleq H_{i,j}^k(n)$ for $1 \leq k \leq K$. Using the notation $w = e^{-j2\pi\Delta f}$, $H_{i,j}$ can be decomposed as

$$H_{i,j} = (I_K \otimes W) A_{i,j} \tag{4.7}$$

where

$$W = \begin{bmatrix} 1 & 1 & \cdots & 1 \\ w^{\tau_0} & w^{\tau_1} & \cdots & w^{\tau_{L-1}} \\ \vdots & \vdots & \ddots & \vdots \\ w^{(N-1)\tau_0} & w^{(N-1)\tau_1} & \cdots & w^{(N-1)\tau_{L-1}} \end{bmatrix}$$

which is related to the delay distribution, and

$$A_{i,j} = \left[\alpha_{i,j}^1(0) \; \alpha_{i,j}^1(1) \; \cdots \; \alpha_{i,j}^1(L-1) \; \cdots \; \alpha_{i,j}^K(0) \; \alpha_{i,j}^K(1) \; \cdots \; \alpha_{i,j}^K(L-1) \right]^T$$

which is related to the power distribution of the channel impulse response. In general, W is not a unitary matrix. If all of the L delay paths fall at the sampling instances of the receiver, W is part of the DFT-matrix, which is unitary. From Equation 4.7, the correlation matrix of the channel frequency response vector between transmit antenna i and receive antenna j can be calculated as

$$R_{i,j} = E\left\{ H_{i,j}H_{i,j}^{\mathcal{H}} \right\} = (I_K \otimes W)E\left\{ A_{i,j}A_{i,j}^{\mathcal{H}} \right\} (I_K \otimes W^{\mathcal{H}})$$

We assume that the MIMO channel is spatially uncorrelated, i.e., the channel coefficients $\alpha_{i,j}^k(l)$ are independent for different indices (i,j). So we can define the time correlation at lag m as $r_T(m) = E\left\{ \alpha_{i,j}^k(l)\alpha_{i,j}^{k+m^*}(l) \right\}$. Thus, the correlation matrix $E\left\{ A_{i,j}A_{i,j}^{\mathcal{H}} \right\}$ can be expressed as

$$E\left\{ A_{i,j}A_{i,j}^{\mathcal{H}} \right\} = R_T \otimes \Lambda \tag{4.8}$$

where $\Lambda = \mathrm{diag}\{\delta_0^2, \delta_1^2, \ldots, \delta_{L-1}^2\}$, and R_T is the temporal correlation matrix of size $K \times K$. We can also define the frequency correlation matrix, R_F, as $R_F = E\{H_{i,j}^k H_{i,j}^k{}^{\mathcal{H}}\}$, where $H_{i,j}^k = [H_{i,j}^k(0), \ldots, H_{i,j}^k(N-1)]^T$. Then, $R_F = W\Lambda W^{\mathcal{H}}$. As a result, we arrive at

$$R_{i,j} = R_T \otimes (W\Lambda W^{\mathcal{H}}) = R_T \otimes R_F \overset{\Delta}{=} R \tag{4.9}$$

where the correlation matrix R is independent of the transceiver antenna indices i and j.

For two distinct STF codewords C and $\tilde{C}$, we denote

$$\Delta \overset{\Delta}{=} (C - \tilde{C})(C - \tilde{C})^{\mathcal{H}} \tag{4.10}$$

Then, the pairwise error probability between C and $\tilde{C}$ can be upper bounded as [26]

$$P(C \to \tilde{C}) \leq \binom{2\nu M_r - 1}{\nu M_r} \left(\prod_{i=1}^{\nu} \lambda_i \right)^{-M_r} \left(\frac{\rho}{M_t} \right)^{-\nu M_r} \tag{4.11}$$

where ν is the rank of $\Delta \circ R$; $\lambda_1, \lambda_2, \ldots, \lambda_\nu$ the nonzero eigenvalues of $\Delta \circ R$; and $\circ$ denotes the Hadamard product.* The minimum value of the

---

* Suppose that $A = \{a_{i,j}\}$ and $B = \{b_{i,j}\}$ are two matrices of size $m \times n$. The *Hadamard product* of A and B is defined as $A \circ B = \{a_{i,j}b_{i,j}\}_{1 \leq i \leq m,\ 1 \leq j \leq n}$.

product $\prod_{i=1}^{\nu} \lambda_i$ over all pairs of distinct signals C and $\tilde{C}$ is termed as *coding advantage*, denoted by

$$\zeta_{\text{STF}} = \min_{C \neq \tilde{C}} \prod_{i=1}^{\nu} \lambda_i \qquad (4.12)$$

Based on the performance upper bound, two STF code design criteria were porposed in Ref. 26.

- *Diversity (rank) criterion*: The minimum rank of $\Delta \circ R$ over all pairs of distinct codewords C and $\tilde{C}$ should be as large as possible.
- *Product criterion*: The coding advantage or the minimum value of the product $\prod_{i=1}^{\nu} \lambda_i$ over all pairs of distinct signals C and $\tilde{C}$ should also be maximized.

If the minimum rank of $\Delta \circ R$ is ν for any pair of distinct STF codewords C and $\tilde{C}$, we say that the STF code achieves a diversity order of νM_r. For a fixed number of OFDM blocks K, number of transmit antennas M_t, and correlation matrices R_T and R_F, the *maximum achievable diversity* or *full diversity* is defined as the maximum diversity order that can be achieved by STF codes of size $KN \times M_t$.

According to the rank inequalities on Hadamard products [37], we have

$$\text{rank}(\Delta \circ R) \leq \text{rank}(\Delta)\text{rank}(R_T)\text{rank}(R_F)$$

Since the rank of Δ is at most M_t and the rank of R_F is at most L, we obtain

$$\text{rank}(\Delta \circ R) \leq \min\{LM_t\text{rank}(R_T), KN\} \qquad (4.13)$$

Thus, the maximum achievable diversity is at most $\min\{LM_tM_r\text{rank}(R_T), KNM_r\}$ in agreement with the results of Ref. 22. However, there is no discussion in Ref. 22 on whether this upper bound can be achieved or not. As we will see later, this upper bound can indeed be achieved. We also observe that if the channel stays constant over multiple OFDM blocks ($\text{rank}(R_T) = 1$), the maximum achievable diversity is only $\min\{LM_tM_r, KNM_r\}$. In this case, STF coding cannot provide additional diversity advantage compared to the SF coding approach.

Note that the above analytical framework includes ST and SF codes as special cases. If we consider only one subcarrier ($N = 1$) and one delay path ($L = 1$), then the channel becomes a single-carrier, time-correlated, flat-fading MIMO channel. The correlation matrix R simplifies to $R = R_T$, and the code design problem reduces to that of ST code design, as described in Ref. 27. In the case of coding over a single OFDM block ($K = 1$), the correlation matrix R becomes $R = R_F$, and the code design problem simplifies to that of SF codes, as discussed in Refs. 19 and 20.

4.3 Full-Diversity SF Codes Design

We introduce in this section two systematic SF code design methods, where coding is applied within each OFDM block. The first method is to obtain full-diversity SF codes from ST codes via mapping [19], which shows that by using a simple repetition mapping, full-diversity SF codes can be constructed from *any* ST (block or trellis) code designed for quasi-static flat Rayleigh fading channels. The other method is to design spectrally efficient SF codes that can guarantee full rate and full diversity for MIMO-OFDM systems with arbitrary power delay profiles [20]. Note that in case of SF coding ($K = 1$), each SF codeword C in Equation 4.2 has a size of $N \times M_t$ and the correlation matrix $R = R_F$ has a size of $N \times N$.

4.3.1 Obtaining Full-Diversity SF Codes from ST Codes via Mapping

For any given integer l ($1 \leq l \leq L$), assume that $lM_t \leq N$ (the number of OFDM subcarriers N is generally larger than LM_t) and k is the largest integer such that $klM_t \leq N$. Suppose that there is a ST encoder with output matrix G. (For ST block encoder, G is a concatenation of some block codewords. For ST trellis encoder, G corresponds to a path of length kM_t starting and ending at the zero state.) Then, a full-diversity SF code C of size $N \times M_t$ can be obtained by mapping the ST codeword G as follows:

$$C = \begin{bmatrix} \mathcal{M}_l(G) \\ \mathbf{0}_{(N-klM_t) \times M_t} \end{bmatrix} \qquad (4.14)$$

where

$$\mathcal{M}_l(G) = [I_{kM_t} \otimes \mathbf{1}_{l \times 1}]G \qquad (4.15)$$

in which $\mathbf{1}_{l \times 1}$ is an all one matrix of size $l \times 1$. Actually, the resulting SF code C is obtained by repeating each row of G l times and adding some zeros. The zero padding used here ensures that the SF code C has size $N \times M_t$, and typically the size of the zero padding is small. The following theorem states that if the employed ST code G has full diversity for flat-fading channels, the SF code constructed by Equation 4.14 will achieve a diversity of at least lM_tM_r [19].

THEOREM 4.1
Suppose that the frequency-selective channel has L independent paths and the maximum path delay is less than one OFDM block period. If an ST (block or trellis) code designed for M_t transmit antennas achieves full diversity for quasi-static flat-fading channels, then the SF code obtained from this ST code via the mapping $\mathcal{M}_l$ ($1 \leq l \leq L$) defined in Equation 4.15 will achieve a diversity order of at least $\min\{lM_tM_r, NM_r\}$.

Moreover, the SF code obtained from an ST block code of square size via the mapping $\mathcal{M}_l$ $(1 \leq l \leq L)$ achieves a diversity of lM_tM_r exactly. Since the maximum achievable diversity is upper bounded by $\min\{LM_tM_r, NM_r\}$; therefore, according to Theorem 4.1, the SF code obtained from a full-diversity ST code via the mapping $\mathcal{M}_L$ defined in Equation 4.15 achieves the maximum achievable diversity $\min\{LM_tM_r, NM_r\}$. We can see that the coding rate of the resulting full-diversity SF codes obtained via the mapping $\mathcal{M}_l$ (Equation 4.15) is $1/l$ times that of the corresponding ST codes, which, however, is larger than that in Ref. 18. For example, for a system with two transmit antennas, eight subcarriers, and a two-ray delay profile, the coding rate of the full-diversity SF codes introduced here is $1/2$, while the coding rate in Ref. 18 is only $1/4$. Note that the simple repetition mapping is independent of particular ST codes, so all the existing ST block and trellis codes achieving full spatial diversity in quasi-static flat-fading environment can be used to design full-diversity SF codes for MIMO-OFDM systems.

To the end of this subsection, we characterize the coding advantage of the resulting SF codes in terms of the coding advantage of the underlying ST codes. We also analyze the effect of the delay distribution and the power distribution on the performance of the proposed SF codes. The *coding advantage* or *diversity product* of a full-diversity ST code for quasi-static flat-fading channels has been defined as [9,27] $\zeta_{ST} = \min_{G \neq \tilde{G}} \left| \prod_{i=1}^{M_t} \beta_i \right|^{1/2M_t}$, where $\beta_1, \beta_2, \ldots,$ β_{M_t} are the nonzero eigenvalues of $(G - \tilde{G})(G - \tilde{G})^{\mathcal{H}}$ for any pair of distinct ST codewords G and $\tilde{G}$. We have the following result [19].

THEOREM 4.2
The diversity product of the full-diversity SF code is bounded by that of the corresponding ST code as follows:

$$\sqrt{\eta_L} \; \Phi \; \zeta_{ST} \leq \zeta_{SF} \leq \sqrt{\eta_1} \; \Phi \zeta_{ST} \tag{4.16}$$

where $\Phi = (\prod_{l=0}^{L-1} \delta_l)^{1/L}$, *and* η_1 *and* η_L *are the largest and smallest eigenvalues, respectively, of the matrix H defined as*

$$H = \begin{bmatrix} H(0) & H(1)^* & \cdots & H(L-1)^* \\ H(1) & H(0) & \cdots & H(L-2)^* \\ \vdots & \vdots & \ddots & \vdots \\ H(L-1) & H(L-2) & \cdots & H(0) \end{bmatrix}_{L \times L} \tag{4.17}$$

and the entries of H are given by $H(n) = \sum_{l=0}^{L-1} e^{-j2\pi n \Delta f \tau_l}$ *for* $n = 0, 1, \ldots, L-1$.

From Theorem 4.2, we can see that the larger the coding advantage of the ST code, the larger the coding advantage of the resulting SF code, suggesting that to maximize the performance of the SF codes, we should look for the best-known ST codes existing in the literature. Moreover, the coding advantage of the SF code depends on the power delay profile. First, it depends on the power distribution through the square root of the geometric average of path powers, i.e., $\Phi = (\prod_{l=0}^{L-1} \delta_l)^{1/L}$. Since the sum of the powers of the paths is unity, this implies that the best performance is expected in case of uniform power distribution (i.e., $\delta_l^2 = 1/L$). Second, the entries of the matrix H defined in Equation 4.17 are functions of the path delays, so the coding advantage also depends on the delay distribution of the paths.

4.3.2 Full-Rate and Full-Diversity SF Code Design

In this subsection, we describe a systematic method to obtain full-rate SF codes achieving full diversity [20]. Specifically, we design a class of SF codes that can achieve a diversity order of $\Gamma M_t M_r$ for any fixed integer Γ $(1 \leq \Gamma \leq L)$.

We consider a coding strategy where each SF codeword C is a concatenation of some matrices G_p:

$$C = \begin{bmatrix} G_1^T & G_2^T & \cdots & G_P^T & \mathbf{0}_{N-P\Gamma M_t}^T \end{bmatrix}^T \quad (4.18)$$

where $P = \lfloor N/(\Gamma M_t) \rfloor$, and each matrix G_p, $p = 1, 2, \ldots, P$, is of size ΓM_t by M_t. The zero padding in Equation 4.18 is used if the number of subcarriers N is not an integer multiple of ΓM_t. Each matrix $G_p (1 \leq p \leq P)$ has the same structure given by

$$G = \sqrt{M_t} \, \mathrm{diag}(X_1, X_2, \ldots, X_{M_t}) \quad (4.19)$$

where $\mathrm{diag}(X_1, X_2, \ldots, X_{M_t})$ is a block diagonal matrix, $X_i = [x_{(i-1)\Gamma+1} \ x_{(i-1)\Gamma+2} \ \cdots \ x_{i\Gamma}]^T$, $i = 1, 2, \ldots, M_t$, and all x_k, $k = 1, 2, \ldots, \Gamma M_t$, are complex symbols and will be specified later. The energy constraint is $E\left(\sum_{k=1}^{\Gamma M_t} |x_k|^2\right) = \Gamma M_t$. For a fixed p, the symbols in G_p are designed jointly, but the designs of G_{p_1} and G_{p_2}, $p_1 \neq p_2$, are independent of each other. The symbol rate of the code is $P\Gamma M_t/N$, ignoring the cyclic prefix. If N is a multiple of ΓM_t, the symbol rate is 1. If not, the rate is less than 1, but since usually N is much greater than ΓM_t, the symbol rate is very close to 1. We have the following sufficient conditions for the SF codes described above to achieve a diversity order of $\Gamma M_t M_r$ [20].

THEOREM 4.3
For any SF code constructed by Equations 4.18 and 4.19, if $\prod_{k=1}^{\Gamma M_t} |x_k - \tilde{x}_k| \neq 0$ for any pair of distinct sets of symbols $\mathbf{X} = [x_1 \ x_2 \ \cdots \ x_{\Gamma M_t}]$ and $\tilde{\mathbf{X}} = [\tilde{x}_1 \ \tilde{x}_2 \ \cdots \ \tilde{x}_{\Gamma M_t}]$, then the SF code achieves a diversity order of $\Gamma M_t M_r$, and the diversity product is

$$\zeta_{SF} = \zeta_{in} \, |\det(Q_0)|^{\frac{1}{2\Gamma}} \quad (4.20)$$

where ζ_{in} is the intrinsic diversity product of the SF code defined as

$$\zeta_{in} = \frac{1}{2} \min_{\mathbf{x} \neq \tilde{\mathbf{x}}} \left(\prod_{k=1}^{\Gamma M_t} |x_k - \tilde{x}_k| \right)^{\frac{1}{\Gamma M_t}} \tag{4.21}$$

and $Q_0 = W_0 \operatorname{diag}(\delta_0^2, \delta_1^2, \ldots, \delta_{L-1}^2) W_0^{\mathcal{H}}$, in which

$$W_0 = \begin{bmatrix} 1 & 1 & \cdots & 1 \\ w^{\tau_0} & w^{\tau_1} & \cdots & w^{\tau_{L-1}} \\ \vdots & \vdots & \ddots & \vdots \\ w^{(\Gamma-1)\tau_0} & w^{(\Gamma-1)\tau_1} & \cdots & w^{(\Gamma-1)\tau_{L-1}} \end{bmatrix}_{\Gamma \times L}$$

From Theorem 4.3, we observe that $|\det(Q_0)|$ depends only on the power delay profile of the channel, and the intrinsic diversity product ζ_{in} depends only on $\min_{\mathbf{x} \neq \tilde{\mathbf{x}}} \left(\prod_{k=1}^{\Gamma M_t} |x_k - \tilde{x}_k| \right)^{1/(\Gamma M_t)}$, which is called the *minimum product distance* of the set of symbols $\mathbf{X} = [x_1 \ x_2 \ldots x_{\Gamma M_t}]$ [28,29]. Therefore, given the code structure Equation 4.38, it is desirable to design the set of symbols $\mathbf{X}$ such that the minimum product distance is as large as possible, a problem that leads to design signal constellations for Rayleigh fading channels [30,31]. A detailed review of the signal design can be found in Ref. 20.

In the sequel, we would like to maximize the coding advantage of the proposed full-rate full-diversity SF codes by permutations. Note that if the transmitter has no *a priori* knowledge about the channel, the performance of the SF codes can be improved by random interleaving, as it can reduce the correlation between adjacent subcarriers. However, if the power delay profile of the channel is available at the transmitter side, further improvement can be achieved by developing a permutation (or interleaving) method that explicitly takes the power delay profile into account. In the following, we assume that the power delay profile of the channel is known at the transmitter. Our objective is to develop an optimum permutation method such that the resulting coding advantage is maximized [20].

THEOREM 4.4
For any subcarrier permutation, the diversity product of the resulting SF code based on Equations 4.18 and 4.19 is

$$\zeta_{SF} = \zeta_{in} \cdot \zeta_{ex} \tag{4.22}$$

where ζ_{in} is the intrinsic diversity products defined in Equation 4.21 and ζ_{ex} is the extrinsic diversity products defined as

$$\zeta_{ex} = \left(\prod_{m=1}^{M_t} \left| \det(V_m \Lambda V_m^{\mathcal{H}}) \right| \right)^{\frac{1}{2\Gamma M_t}} \tag{4.23}$$

in which $\Lambda = \text{diag}(\delta_0^2, \delta_1^2, \ldots, \delta_{L-1}^2)$ *and*

$$
V_m = \begin{bmatrix}
1 & 1 & \cdots & 1 \\
w^{[n_{(m-1)\Gamma+2}-n_{(m-1)\Gamma+1}]\tau_0} & w^{[n_{(m-1)\Gamma+2}-n_{(m-1)\Gamma+1}]\tau_1} & \cdots & w^{[n_{(m-1)\Gamma+2}-n_{(m-1)\Gamma+1}]\tau_{L-1}} \\
\vdots & \vdots & \ddots & \vdots \\
w^{[n_{m\Gamma}-n_{(m-1)\Gamma+1}]\tau_0} & w^{[n_{m\Gamma}-n_{(m-1)\Gamma+1}]\tau_1} & \cdots & w^{[n_{m\Gamma}-n_{(m-1)\Gamma+1}]\tau_{L-1}}
\end{bmatrix}
$$

(4.24)

Moreover, the extrinsic diversity product ζ_{ex} *is upper bounded as*

(i) $\zeta_{\text{ex}} \leq 1$
and more precisely,

(ii) *if we sort the power profile* $\delta_0, \delta_1, \ldots, \delta_{L-1}$ *in a nonincreasing order as*
$\delta_{l_1} \geq \delta_{l_2} \geq \cdots \geq \delta_{l_L}$, *then*

$$
\zeta_{\text{ex}} \leq \left(\prod_{i=1}^{\Gamma} \delta_{l_i} \right)^{\frac{1}{\Gamma}} \left| \prod_{m=1}^{M_t} \det(V_m V_m^{\mathcal{H}}) \right|^{\frac{1}{2\Gamma M_t}}
$$

(4.25)

where equality holds when $\Gamma = L$. *As a consequence,* $\zeta_{\text{ex}} \leq \sqrt{L}(\prod_{i=1}^{\Gamma} \delta_{l_i})^{1/\Gamma}$.

We observe that the extrinsic diversity product ζ_{ex} depends on the power delay profile in two ways. First, it depends on the power distribution through the square root of the geometric average of the largest Γ path powers, i.e., $(\prod_{i=1}^{\Gamma} \delta_{l_i})^{1/\Gamma}$. In case of $\Gamma = L$, the best performance is expected if the power distribution is uniform (i.e., $\delta_l^2 = 1/L$) since the sum of the path powers is unity. Second, the extrinsic diversity product ζ_{ex} also depends on the delay distribution and the applied subcarrier permutation. In contrast, the intrinsic diversity product, ζ_{in}, is not affected by the power delay profile or the permutation method, and it depends only on the signal constellation and the SF code design.

By carefully choosing the applied permutation method, the overall performance of the SF code can be improved by increasing the value of the extrinsic diversity product ζ_{ex}. Toward this end, we consider a specific permutation strategy as follows. We decompose any integer n ($0 \leq n \leq N - 1$) as

$$
n = e_1 \Gamma + e_0
$$

(4.26)

where $0 \leq e_0 \leq \Gamma - 1$, $e_1 = \lfloor \frac{n}{\Gamma} \rfloor$, and $\lfloor x \rfloor$ denotes the largest integer not greater than x. For a fixed integer μ ($\mu \geq 1$), we further decompose e_1 in Equation 4.26 as

$$
e_1 = v_1 \mu + v_0
$$

(4.27)

where $0 \leq v_0 \leq \mu - 1$ and $v_1 = \lfloor \frac{e_1}{\mu} \rfloor$. We permute the rows of the $N \times M_t$ SF codeword constructed from Equations 4.37 and 4.38 in such a way that the

nth $(0 \leq n \leq N-1)$ row of C is moved to the $\sigma(n)$th row, where

$$\sigma(n) = v_1 \mu \Gamma + e_0 \mu + v_0 \qquad (4.28)$$

in which e_0, v_0, and v_1 come from Equations 4.26 and 4.27. We call the integer μ as the *separation factor*. The separation factor μ should be chosen such that $\sigma(n) \leq N$ for any $0 \leq n \leq N-1$, or equivalently, $\mu \leq \lfloor N/\Gamma \rfloor$. Moreover, to guarantee that the mapping Equation 4.28 is one-to-one over the set $\{0, 1, \ldots, N-1\}$ (i.e., it defines a permutation), μ must be a factor of N. The role of the permutation specified in Equation 4.28 is to separate two neighboring rows of C by μ subcarriers. The following result characterizes the extrinsic diversity product of the SF code that is permuted with the above-described method [20].

THEOREM 4.5
For the permutation specified in Equation 4.28 with a separation factor μ, the extrinsic diversity product of the permuted SF code is

$$\zeta_{ex} = \left| \det(V_0 \Lambda V_0^{\mathcal{H}}) \right|^{\frac{1}{2\Gamma}} \qquad (4.29)$$

where

$$V_0 = \begin{bmatrix} 1 & 1 & \cdots & 1 \\ w^{\mu \tau_0} & w^{\mu \tau_1} & \cdots & w^{\mu \tau_{L-1}} \\ w^{2\mu \tau_0} & w^{2\mu \tau_1} & \cdots & w^{2\mu \tau_{L-1}} \\ \vdots & \vdots & \ddots & \vdots \\ w^{(\Gamma-1)\mu \tau_0} & w^{(\Gamma-1)\mu \tau_1} & \cdots & w^{(\Gamma-1)\mu \tau_{L-1}} \end{bmatrix}_{\Gamma \times L} \qquad (4.30)$$

Moreover, if $\Gamma = L$, the extrinsic diversity product ζ_{ex} can be calculated as

$$\zeta_{ex} = \left(\prod_{l=0}^{L-1} \delta_l \right)^{\frac{1}{L}} \left\{ \prod_{0 \leq l_1 < l_2 \leq L-1} \left| 2 \sin \left[\frac{\mu(\tau_{l_2} - \tau_{l_1})\pi}{T} \right] \right| \right\}^{\frac{1}{L}} \qquad (4.31)$$

The permutation (Equation 4.28) is determined by the separation factor μ. Our objective is to find a separation factor μ_{op} that maximizes the extrinsic diversity product ζ_{ex}, i.e., $\mu_{op} = \arg \max_{1 \leq \mu \leq \lfloor N/\Gamma \rfloor} |\det(V_0 \Lambda V_0^{\mathcal{H}})|$. If $\Gamma = L$, the optimum separation factor μ_{op} can be expressed as

$$\mu_{op} = \arg \max_{1 \leq \mu \leq \lfloor N/\Gamma \rfloor} \prod_{0 \leq l_1 < l_2 \leq L-1} \left| \sin \left[\frac{\mu(\tau_{l_2} - \tau_{l_1})\pi}{T} \right] \right| \qquad (4.32)$$

which is independent of the path powers. The optimum separation factor can be easily found via low complexity computer search. However, in some cases, closed form solutions can also be obtained:

- If $\Gamma = L = 2$, the extrinsic diversity product ζ_{ex} is

$$\zeta_{ex} = \sqrt{\delta_0 \delta_1} \left| 2 \sin \left[\frac{\mu(\tau_1 - \tau_0)\pi}{T} \right] \right|^{\frac{1}{2}} \qquad (4.33)$$

Suppose that the system has $N = 128$ subcarriers, and the total bandwidth is $BW = 1\,MHz$. Then, the OFDM block duration is $T = 128\,\mu s$ without the cyclic prefix. If $\tau_1 - \tau_0 = 5\,\mu s$, then $\mu_{op} = 64$ and $\zeta_{ex} = \sqrt{2\delta_0 \delta_1}$. If $\tau_1 - \tau_0 = 20\,\mu s$, then $\mu_{op} = 16$ and $\zeta_{ex} = \sqrt{2\delta_0 \delta_1}$. In general, if $\tau_1 - \tau_0 = 2^a b\,\mu s$, where a is a nonnegative integer and b is an odd integer, $\mu_{op} = 128/2^{a+1}$. In all of these cases, the extrinsic diversity product is $\zeta_{ex} = \sqrt{2\delta_0 \delta_1}$, which achieves the upper bound in Theorem 4.4.

- Assume that $\tau_l - \tau_0 = l N_0 T/N$, $l = 1, 2, \ldots, L-1$, and N is an integer multiple of LN_0, where N_0 is a constant and not necessarily an integer. If $\Gamma = L$ or $\delta_0^2 = \delta_1^2 = \cdots = \delta_{L-1}^2 = 1/L$, the optimum separation factor is

$$\mu_{op} = \frac{N}{LN_0} \qquad (4.34)$$

and the corresponding extrinsic diversity product is $\zeta_{ex} = \sqrt{L} \left(\prod_{l=0}^{L-1} \delta_l \right)^{1/L}$. In particular, in case of $\delta_0^2 = \delta_1^2 = \cdots = \delta_{L-1}^2 = 1/L$, $\zeta_{ex} = 1$. In both cases, the extrinsic diversity products achieve the upper bounds of Theorem 4.4. Note that if $\tau_l = lT/N$ for $l = 0, 1, \ldots, L-1$, $\Gamma = L$ and N is an integer multiple of L, the permutation with the optimum separation factor $\mu_{op} = N/L$ is similar to the optimum subcarrier grouping method proposed in Ref. 25, which is not optimal for arbitrary power delay profiles.

4.4 Full-Diversity STF Code Design

In this section, we review two STF code design methods to achieve the maximum achievable diversity order $\min\{LM_t M_r \text{rank}(R_T), KNM_r\}$ [26]. Assume that the number of subcarriers N is not less than LM_t, so the maximum achievable diversity order is $LM_t M_r \text{rank}(R_T)$.

4.4.1 Repetition-Coded STF Code Design

First, we try to systematically design full-diversity STF codes by taking advantage of the full-diversity SF codes discussed in the previous section.

Specifically, assume that C_{SF} is a full-diversity SF code of size $N \times M_t$, then we can construct a full-diversity STF code, C_{STF}, by repeating C_{SF} K times (over K OFDM blocks) as follows:

$$C_{STF} = \mathbf{1}_{k \times 1} \otimes C_{SF} \tag{4.35}$$

where $\mathbf{1}_{k \times 1}$ is an all one matrix of size $k \times 1$. Denote

$$\Delta_{STF} = (C_{STF} - \tilde{C}_{STF})(C_{STF} - \tilde{C}_{STF})^{\mathcal{H}}$$

and

$$\Delta_{SF} = (C_{SF} - \tilde{C}_{SF})(C_{SF} - \tilde{C}_{SF})^{\mathcal{H}}$$

Then, we have

$$\Delta_{STF} = \left[\mathbf{1}_{k \times 1} \otimes (C_{SF} - \tilde{C}_{SF}) \right] \left[\mathbf{1}_{1 \times k} \otimes (C_{SF} - \tilde{C}_{SF})^{\mathcal{H}} \right] = \mathbf{1}_{k \times k} \otimes \Delta_{SF}$$

Thus,

$$\Delta_{STF} \circ R = (\mathbf{1}_{k \times k} \otimes \Delta_{SF}) \circ (R_T \otimes R_F) = R_T \otimes (\Lambda_{SF} \circ R_F)$$

Since the SF code C_{SF} achieves full diversity in each OFDM block, the rank of $\Delta_{SF} \circ R_F$ is LM_t. Therefore, the rank of $\Delta_{STF} \circ R$ is $LM_t\mathrm{rank}(R_T)$, so C_{STF} in Equation 4.35 is guaranteed to achieve a diversity order of $LM_tM_r\mathrm{rank}(R_T)$.

We observe that the maximum achievable diversity depends on the rank of the temporal correlation matrix R_T. If the fading channels are constant during K OFDM blocks, i.e., $\mathrm{rank}(R_T) = 1$, then the maximum achievable diversity order for STF codes (coding across several OFDM blocks) is the same as that for SF codes (coding within one OFDM block). Moreover, if the channel changes independently in time, i.e., $R_T = I_K$, the repetition structure of STF code C_{STF} in Equation 4.35 is sufficient, but not necessary, to achieve the full diversity. In this case,

$$\Delta \circ R = \mathrm{diag}(\Delta_1 \circ R_F, \Delta_2 \circ R_F, \ldots, \Delta_K \circ R_F)$$

where $\Delta_k = (C_k - \tilde{C}_k)(C_k - \tilde{C}_k)^{\mathcal{H}}$ for $1 \leq k \leq K$. Thus, in this case, the necessary and sufficient condition to achieve full diversity KLM_tM_r is that each matrix $\Delta_k \circ R_F$ be of rank LM_t over all pairs of distinct codewords simultaneously for all $1 \leq k \leq K$.

Note that the above repetition-coded STF code design ensures full diversity at the price of symbol rate decrease by a factor of $1/K$ (over K OFDM blocks) compared to the symbol rate of the underlying SF code. The advantage of this approach is that any full-diversity SF code (block or trellis) can be used to design full-diversity STF codes.

4.4.2   Full-Rate Full-Diversity STF Code Design

Second, we would like to design a class of STF codes that can guarantee a diversity order of $\Gamma M_t M_r \mathrm{rank}(R_T)$ for any given integer $\Gamma(1 \leq \Gamma \leq L)$ by extending the full-rate full-diversity SF code construction method (coding over one OFDM block, i.e., the $K=1$ case) as discussed in the previous section.

We consider an STF code structure consisting of STF codewords C of size KN by M_t:

$$C = \begin{bmatrix} C_1^T & C_2^T & \cdots & C_K^T \end{bmatrix}^T \tag{4.36}$$

where

$$C_k = \begin{bmatrix} G_{k,1}^T & G_{k,2}^T & \cdots & G_{k,P}^T & 0_{N-P\Gamma M_t}^T \end{bmatrix}^T \tag{4.37}$$

for $k=1, 2, \ldots, K$. In Equation 4.37, $P=\lfloor N/(\Gamma M_t)\rfloor$, and each matrix $G_{k,p} (1 \leq k \leq K, 1 \leq p \leq P)$ is of size ΓM_t by M_t. The zero padding in Equation 4.37 is used if the number of subcarriers N is not an integer multiple of ΓM_t. For each $p(1 \leq p \leq P)$, we design the code matrices $G_{1,p}, G_{2,p}, \ldots, G_{K,p}$ jointly, but the designs of G_{k_1,p_1} and G_{k_2,p_2}, $p_1 \neq p_2$, are independent of each other. For a fixed $p(1 \leq p \leq P)$, let

$$G_{k,p} = \sqrt{M_t}\, \mathrm{diag}(X_{k,1}, X_{k,2}, \ldots, X_{k,M_t}), \quad k=1, 2, \ldots, K \tag{4.38}$$

where $\mathrm{diag}(X_{k,1}, X_{k,2}, \ldots, X_{k,M_t})$ is a block diagonal matrix, $X_{k,i} = [x_{k,(i-1)\Gamma+1}$ $x_{k,(i-1)\Gamma+2} \cdots x_{k,i\Gamma}]^T$, $i=1, 2, \ldots, M_t$, and all $x_{k,j}$, $j=1, 2, \ldots, \Gamma M_t$, are complex symbols and will be specified later. The energy normalization condition is $E\left(\sum_{k=1}^{K}\sum_{j=1}^{\Gamma M_t}|x_{k,j}|^2\right) = K\Gamma M_t$. The symbol rate of the proposed scheme is $P\Gamma M_t/N$, ignoring the cyclic prefix. If N is a multiple of ΓM_t, the symbol rate is 1. If not, the rate is less than 1, but since usually N is much greater than ΓM_t, the symbol rate is very close to 1.

The following theorem provides a sufficient condition for the STF codes described above to achieve a diversity order of $\Gamma M_t M_r \mathrm{rank}(R_T)$. For simplicity, we use the notation $X=[x_{1,1} \cdots x_{1,\Gamma M_t} \cdots x_{K,1} \cdots x_{K,\Gamma M_t}]$ and $\tilde{X}=[\tilde{x}_{1,1} \cdots \tilde{x}_{1,\Gamma M_t} \cdots \tilde{x}_{K,1} \cdots \tilde{x}_{K,\Gamma M_t}]$. Moreover, for any $n \times n$ nonnegative definite matrix A, we denote its eigenvalues in a nonincreasing order as: $\mathrm{eig}_1(A) \geq \mathrm{eig}_2(A) \geq \cdots \geq \mathrm{eig}_n(A)$.

THEOREM 4.6
For any STF code constructed by Equations 4.36 through 4.38, if $\prod_{k=1}^{K}\prod_{j=1}^{\Gamma M_t} \times$
$|x_{k,j} - \tilde{x}_{k,j}| \neq 0$ for any pair of distinct symbols X and $\tilde{X}$, then the STF code achieves a diversity order of $\Gamma M_t M_r \mathrm{rank}(R_T)$, and the coding advantage is bounded by

$$(M_t\delta_{\min})^{\Gamma M_t \mathrm{rank}(R_T)}\, \Phi \leq \zeta_{STF} \leq (M_t\delta_{\max})^{\Gamma M_t \mathrm{rank}(R_T)}\, \Phi \tag{4.39}$$

where

$$\delta_{\min} = \min_{X \neq \tilde{X}} \min_{1 \leq k \leq K, 1 \leq j \leq \Gamma M_t} |x_{k,j} - \tilde{x}_{k,j}|^2$$

$$\delta_{\max} = \max_{X \neq \tilde{X}} \max_{1 \leq k \leq K, 1 \leq j \leq \Gamma M_t} |x_{k,j} - \tilde{x}_{k,j}|^2$$

$$\Phi = |\det(Q_0)|^{M_t \text{rank}(R_T)} \prod_{i=1}^{\text{rank}(R_T)} (\text{eig}_i(R_T))^{\Gamma M_t} \tag{4.40}$$

and

$$Q_0 = W_0 \text{diag}\left(\delta_0^2, \delta_1^2, \ldots, \delta_{L-1}^2\right) W_0^{\mathcal{H}} \tag{4.41}$$

$$W_0 = \begin{bmatrix} 1 & 1 & \cdots & 1 \\ w^{\tau_0} & w^{\tau_1} & \cdots & w^{\tau_{L-1}} \\ \vdots & \vdots & \ddots & \vdots \\ w^{(\Gamma-1)\tau_0} & w^{(\Gamma-1)\tau_1} & \cdots & w^{(\Gamma-1)\tau_{L-1}} \end{bmatrix}_{\Gamma \times L} \tag{4.42}$$

Furthermore, if the temporal correlation matrix R_T is of full rank, i.e., $\text{rank}(R_T) = K$, the coding advantage is

$$\zeta_{\text{STF}} = \delta M_t^{K\Gamma M_t} |\det(R_T)|^{\Gamma M_t} |\det(Q_0)|^{K M_t} \tag{4.43}$$

where

$$\delta = \min_{X \neq \tilde{X}} \prod_{k=1}^{K} \prod_{j=1}^{\Gamma M_t} |x_{k,j} - \tilde{x}_{k,j}|^2 \tag{4.44}$$

We observe from Theorem 4.6 that with the code structure specified in Equations 4.36 through 4.38, it is not difficult to achieve the diversity order of $\Gamma M_t M_r \text{rank}(R_T)$ if we have signals satisfying $\prod_{k=1}^{K} \prod_{j=1}^{\Gamma M_t} |x_{k,j} - \tilde{x}_{k,j}| \neq 0$ for distinct symbols X and $\tilde{X}$. However, it is challenging to design a set of complex symbol vectors, $X = [x_{1,1} \cdots x_{1,\Gamma M_t} \cdots x_{K,1} \cdots x_{K,\Gamma M_t}]$, such that the coding advantage ζ_{STF} is as large as possible. One approach is to maximize $\delta_{\min}$ and $\delta_{\max}$ in Equation 4.39 according to the lower and upper bounds of the coding advantage. Another approach is to maximize δ in Equation 4.44. We would like to design signals according to the second criterion for two reasons. First, the coding advantage ζ_{STF} in Equation 4.43 is determined by δ in closed form although this closed form only holds with the assumption that the temporal correlation matrix R_T is of full rank. Second, the problem of designing X to maximize δ is related to the problem of constructing signal constellations for Rayleigh fading channels which have been well developed [28–31]. In the literature, δ is called the *minimum product distance* of the set of symbols X [28,29].

We summarize some existing results on designing X to maximize the minimum product distance δ as follows. For simplicity, denote $\mathcal{L} = K\Gamma M_t$, and assume that Ω is a constellation such as QAM, PAM, and so on. The set of complex symbol vectors is obtained by applying a transform over a $\mathcal{L}$-dimensional signal set $\Omega^{\mathcal{L}}$ [30,31]. Specifically,

$$\mathbf{X} = S \cdot \frac{1}{\sqrt{\mathcal{L}}} V(\theta_1, \theta_2, \ldots, \theta_{\mathcal{L}}) \tag{4.45}$$

where $S = [s_1 \, s_2 \, \cdots \, s_{\mathcal{L}}] \in \Omega^K$ is a vector of arbitrary channel symbols to be transmitted, and $V(\theta_1, \theta_2, \ldots, \theta_{\mathcal{L}})$ a Vandermonde matrix with variables $\theta_1, \theta_2, \ldots, \theta_{\mathcal{L}}$ [37]:

$$V(\theta_1, \theta_2, \ldots, \theta_{\mathcal{L}}) = \begin{bmatrix} 1 & 1 & \cdots & 1 \\ \theta_1 & \theta_2 & \cdots & \theta_{\mathcal{L}} \\ \vdots & \vdots & \ddots & \vdots \\ \theta_1^{\mathcal{L}-1} & \theta_2^{\mathcal{L}-1} & \cdots & \theta_{\mathcal{L}}^{\mathcal{L}-1} \end{bmatrix} \tag{4.46}$$

The optimum θ_l, $1 \leq l \leq \mathcal{L}$, has been specified for different $\mathcal{L}$ and Ω. For example, if Ω is a QAM constellation and $\mathcal{L} = 2^s (s \geq 1)$, the optimum θ_l values were given by [30,31]

$$\theta_l = e^{j\frac{4l-3}{2\mathcal{L}}\pi}, \quad l = 1, 2, \ldots, \mathcal{L} \tag{4.47}$$

In case of $\mathcal{L} = 2^s \cdot 3^t (s \geq 1, t \geq 1)$, a class of θ_l values were given in Ref. 31 as

$$\theta_l = e^{j\frac{6l-5}{3\mathcal{L}}\pi}, \quad l = 1, 2, \ldots, \mathcal{L} \tag{4.48}$$

For more details and other cases of Ω and $\mathcal{L}$, we refer the reader to Refs. 30 and 31.

The STF code design discussed in this subsection achieves full symbol rate, which is much larger than that of the repetition coding approach. However, the maximum-likelihood decoding complexity of this approach is high. Its complexity increases exponentially with the number of OFDM blocks, K, while the decoding complexity of the repetition-coded STF codes increases only linearly with K. Fortunately, sphere decoding methods [32,33] can be used to reduce the complexity.

4.5   Simulation Results

We simulated MIMO-OFDM systems with the proposed SF/STF code designs and present average BER performance curves as functions of the average SNR. The simulated systems have $M_t = 2$ transmit antennas and $M_r = 1$ receive antenna, and the OFDM modulation has $N = 128$ subcarriers. We considered two fading channel models. The first one is a two-ray, equal-power delay profile with a delay $\tau\,\mu s$ between the two rays and each ray was modeled as a zero-mean, complex Gaussian random variable with variance 0.5. The second channel model is a more realistic six-ray Typical Urban (TU) power delay profile [36] with delay distribution $\{0.0\,\mu s, 0.2\,\mu s, 0.5\,\mu s, 1.6\,\mu s, 2.3\,\mu s, 5.0\,\mu s\}$ and power distribution $\{0.189, 0.379, 0.239, 0.095, 0.061, 0.037\}$.

First, we simulated a full-diversity SF code obtained from ST code via mapping as proposed in Equations 4.14 and 4.15. We used the orthogonal ST block codes for two transmit antennas as follows [7]:

$$G_2 = \begin{bmatrix} x_1 & x_2 \\ -x_2^* & x_1^* \end{bmatrix} \tag{4.49}$$

where x_1 and x_2 are PSK or QAM information symbols. The SF code was obtained by repeating each row of the orthogonal ST code (Equation 4.49) twice. We simulated the SF code over the two-ray fading channel model in which we considered two cases: (a) $\tau = 5\,\mu s$ and (b) $\tau = 20\,\mu s$ with OFDM bandwidth BW $= 1$ MHz. We compared the performances of the full-diversity SF code (with row repetition) with that of the scheme by applying the orthogonal design directly (without row repetition). We used BPSK modulation for the nonrepeated case and QPSK for the repeated case. Therefore, both schemes had a same spectral efficiency of 1 bit/s/Hz. From Figure 4.1, we can see that in case of $\tau = 20\,\mu s$, the performance curve of the full-diversity SF code has a steeper slope than that of the code without repetition, i.e., the obtained SF code has higher diversity order. We can observe a performance improvement of about 4 dB at a BER of 10^{-4}. From Figure 4.2, we observe that the performance of the full-diversity SF code degraded significantly from $\tau = 20\,\mu s$ case to $\tau = 5\,\mu s$ case, which is consistent with the theoretical result that the coding advantage depends on the delay distribution of the multiple paths. Note that the correlation of the channel frequency responses at adjacent subcarriers can severely degrade the performance of the SF codes, so we may apply a random permutation to break down the correlation to improve the code performance. We considered a random permutation generated by the Takeshita–Constello method [34], which is given by

$$\sigma(n) = \left[\frac{n(n+1)}{2}\right] \bmod N, \quad n = 0, 1, \ldots, N - 1 \tag{4.50}$$

From Figures 4.1 and 4.2, we can see that by applying a random permutation over the channel frequency responses, the performance of the full-diversity SF code can be further improved and the performance improvement depends on the channel power delay distribution.

Second, we simulated the full-rate full-diversity SF code design proposed in Equations 4.18 and 4.19. We considered an SF code for $M_t = 2$ transmit

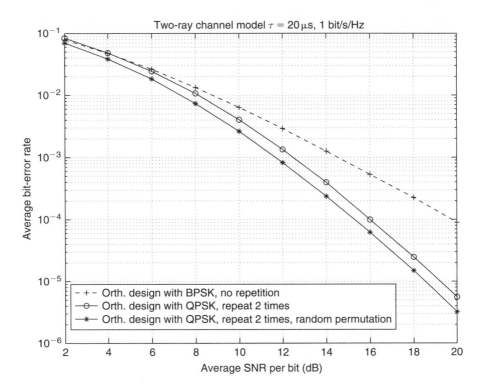

FIGURE 4.1
Performances of the full-diversity SF code obtained from orthogonal ST code via mapping in a two-ray fading model with $\tau = 20\,\mu s$.

antennas with the following structure

$$G = \sqrt{2} \begin{bmatrix} x_1 & 0 \\ x_2 & 0 \\ 0 & x_3 \\ 0 & x_4 \end{bmatrix} \tag{4.51}$$

The symbols x_1, x_2, x_3, x_4 were obtained as $[x_1\ x_2\ x_3\ x_4] = [s_1\ s_2\ s_3\ s_4] \cdot \frac{1}{2} V(\theta, -\theta, j\theta, -j\theta)$, where s_1, s_2, s_3, s_4 were chosen from QPSK constellation ($s_i \in \{\pm 1, \pm j\}$), $V(\cdot)$ is the Vandermonde matrix defined in Equation 4.46, and $\theta = e^{j\pi/8}$. This code targets a frequency diversity order of $\Gamma = 2$, thus it achieves full diversity only if the number of delay paths is $L \le 2$. We simulated the SF code over the more realistic six-ray TU fading model with two scenarios: (a) BW $= 1\,\text{MHz}$, and (b) BW $= 4\,\text{MHz}$. We first compared the SF code performances by using three different permutation schemes: no permutation, random permutation, and the proposed optimum permutation. The random permutation was generated by the Takeshita–Constello method in Equation 4.50. From Figures 4.3 and 4.4, we can see that the performance of the proposed SF code with the

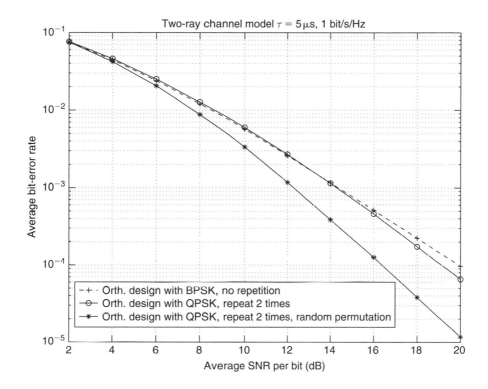

FIGURE 4.2
Performances of the full-diversity SF code obtained from orthogonal ST code via mapping in a two-ray fading model with $\tau = 5\,\mu s$.

random permutation is much better than that without permutation. If we apply the proposed optimum permutation with a separation factor $\mu = 64$, the code performance can further be improved and there is an additional performance gain of 1.5 dB and 1 dB at a BER of 10^{-5} in case of BW $= 1$ MHz and BW $= 4$ MHz, respectively.

We also compared the full-rate full-diversity SF code design (Equation 4.51) with the full-diversity SF code obtained from the orthogonal ST code (Equation 4.49) with repetition. The full-diversity SF code based on Equation 4.49 used 16-QAM modulation to maintain the same spectral efficiency of 2 bits/s/Hz as that of Equation 4.51. We observe from Figure 4.3 that in case of BW $= 1$ MHz, the SF code in Equation 4.51 outperforms the SF code based on repetition by about 2 dB at a BER of 10^{-4}. With random permutation (Equation 4.50), the performance of the full-rate SF code is better than that of the code from orthogonal design by about 2.5 dB at a BER of 10^{-4}. Moreover, we have an additional performance improvement of 1 dB at a BER of 10^{-4} achieved by the full-rate SF code when the optimum permutation strategy is applied. In case of BW $= 4$ MHz, from Figure 4.4, we can see that without permutation, the performance of the SF code in Equation 4.51 is better than

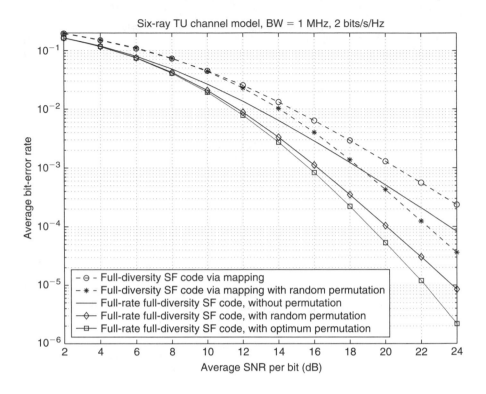

FIGURE 4.3
Performances of the full-diversity SF codes with different permutation strategies in the six-ray
TU fading model, BW = 1 MHz.

that of the code from orthogonal design by about 3 dB at a BER of 10^{-4}. With
random permutation, the SF code (Equation 4.51) outperforms the SF code
based on Equation 4.49 by about 2 dB at a BER of 10^{-4}. Compared to the SF
code from orthogonal design with the random permutation, the full-rate SF
code with the optimum permutation has a total gain of 3 dB at a BER of 10^{-4}.

Finally, we simulated a full-rate full-diversity STF code based on Equations
4.36 through 4.38 for $M_t = 2$ transmit antennas with $\Gamma = 2$ and over $K = 1$,
2, 3 OFDM blocks. The set of complex symbol vectors **X** was obtained via
Equation 4.45 by applying Vandermonde transforms over a signal set Ω^{4K} for
$K = 1, 2, 3$. The Vandermonde transforms were determined for different K values according to Equations 4.47 and 4.48. The constellation Ω was chosen to be
BPSK. Thus, the spectral efficiency the resulting STF codes were 1 bit/s/Hz
(omitting the cyclic prefix), which is independent of the number of jointly
encoded OFDM blocks, K. We considered the six-ray TU channel model and
assumed that the fading channel is constant within each OFDM block period
but varies from one OFDM block period to another according to a first-order
Makovian model [35]: $\alpha_{i,j}^k(l) = \varepsilon \alpha_{i,j}^{k-1}(l) + \eta_{i,j}^k(l)$, $0 \le l \le L - 1$, where the
constant ε ($0 \le \varepsilon \le 1$) determines the amount of temporal correlation, and $\eta_{i,j}^k(l)$

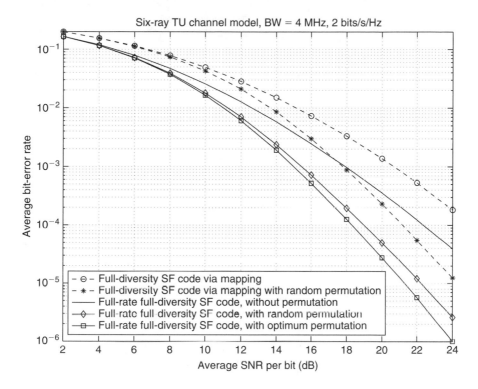

FIGURE 4.4
Performances of the full-diversity SF codes with different permutation strategies in the six-ray TU fading model, BW = 4 MHz.

is a zero-mean, complex Gaussian random variable with variance $\delta_l \sqrt{1 - \varepsilon^2}$. If $\varepsilon = 0$, there is no temporal correlation (independent fading), while if $\varepsilon = 1$, the channel stays constant over multiple OFDM blocks. We considered three temporal correlation scenarios: $\varepsilon = 0$, $\varepsilon = 0.8$, and $\varepsilon = 0.95$. The performance of the full-rate STF codes are depicted in Figure 4.5 for the three different temporal correlation scenarios. From the figures, we observe that the diversity order of the STF codes increases with the number of jointly encoded OFDM blocks, K. However, the improvement of the diversity order depends on the temporal correlation. The performance gain obtained by coding across multiple OFDM blocks decreases as the correlation factor ε increases.

4.6 Conclusion

In this chapter, we reviewed code design criteria for MIMO-OFDM systems and summarized our results on SF and STF code design [19,20,26]. We explored different coding approaches for MIMO-OFDM systems by taking

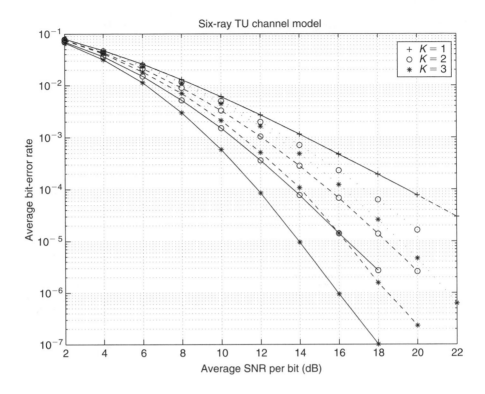

FIGURE 4.5
Performances of the full-rate full-diversity STF codes over $K = 1, 2, 3$ OFDM blocks. Note that the dotted curves show the results for the channel with correlation $\varepsilon = 0.95$, the broken curves for the channel with correlation $\varepsilon = 0.8$, and the solid curve for channel with correlation $\varepsilon = 0$.

into account all opportunities for performance improvement in the spatial, the temporal, and the frequency domains in terms of the achievable diversity order. For SF coding, where the coding is applied within each OFDM block, we proposed two systematic SF code design methods that can guarantee to achieve the full-diversity order of LM_tM_r, where factor L comes from the frequency diversity owing to the delay spread of the channel. For STF coding, where coding is applied over multiple OFDM blocks, we developed two STF code design methods by taking advantage of the SF code design methodology. The proposed STF code design methods can guarantee the maximum achievable diversity order of LM_tM_rT, where T is the rank of the temporal correlation matrix of the fading channel. The simulation results demonstrate that using the proposed coding methods, considerable performance improvement can be achieved compared to previously existing SF and STF coding approaches. We also observed that the performance of the obtained SF and STF codes depend heavily on the channel power delay profile, and there is trade-off between the diversity order and the spectral efficiency of the code.

References

1. Air interface for fixed broadband wireless access systems, *IEEE Standard 802.16-2004*, October 2004.
2. Air interface for fixed and mobile broadband wireless access systems, *IEEE Standard Amendment P802.16e/D12*, February 2005.
3. Mobile WiMAX—Part I: A technical overview and performance evaluation, *WiMAX Forum*, June 2006.
4. A. Ghosh and D. Wolter, Broadband wireless access with WiMax/802.16: Current performance benchmarks and future potential, *IEEE Commun. Magaz.*, pp. 129–136, February 2005.
5. J.-C. Guey, M. P. Fitz, M. R. Bell, and W.-Y. Kuo, Signal design for transmitter diversity wireless communication systems over Rayleigh fading channels, *IEEE Trans. Commun.*, vol. 47, pp. 527–537, April 1999.
6. V. Tarokh, N. Seshadri, and A. R. Calderbank, Space-time codes for high data rate wireless communication: Performance criterion and code construction, *IEEE Trans. Inform. Theory*, vol. 44, no. 2, pp. 744–765, 1998.
7. S. Alamouti, A simple transmit diversity technique for wireless communications, *IEEE JSAC*, vol. 16, no. 8, pp. 1451–1458, 1998.
8. V. Tarokh, H. Jafarkhani, and A. R. Calderbank, Space-time block codes from orthogonal designs, *IEEE Trans. Inform. Theory*, vol. 45, no. 5, pp. 1456–1467, 1999.
9. B. M. Hochwald and T. L. Marzetta, Unitary space-time modulation for multiple-antenna communication in Rayleigh flat fading, *IEEE Trans. Inform. Theory*, vol. 46, no. 2, pp. 543–564, 2000.
10. M. O. Damen, K. Abed-Meraim, and J. C. Belfiore, Diagonal algebraic space-time block codes, *IEEE Trans. Inform. Theory*, vol. 48, no. 3, pp. 628–636, March 2002.
11. D. Agrawal, V. Tarokh, A. Naguib, and N. Seshadri, Space-time coded OFDM for high data-rate wireless communication over wideband channels, *Proc. IEEE VTC*, pp. 2232–2236, 1998.
12. K. Lee and D. Williams, A space-frequency transmitter diversity technique for OFDM systems, *Proc. IEEE GLOBECOM*, vol. 3, pp. 1473–1477, 2000.
13. R. Blum, Y. Li, J. Winters, and Q. Yan, Improved space-time coding for MIMO-OFDM wireless communications, *IEEE Trans. Commun.*, vol. 49, no. 11, pp. 1873–1878, 2001.
14. Y. Gong and K. B. Letaief, An efficient space-frequency coded wideband OFDM system for wireless communications, *Proc. IEEE ICC*, vol. 1, pp. 475–479, 2002.
15. Z. Hong and B. Hughes, Robust space-time codes for broadband OFDM systems, *Proc. IEEE WCNC*, vol. 1, pp. 105–108, 2002.
16. B. Lu and X. Wang, Space-time code design in OFDM systems, *Proc. IEEE GLOBECOM*, pp. 1000–1004, November 2000.
17. H. Bölcskei and A. J. Paulraj, Space-frequency coded broadband OFDM systems, *Proc. IEEE WCNC*, pp. 1–6, September 2000.
18. H. Bölcskei and A. J. Paulraj, Space-frequency codes for broadband fading channels, *Proc. ISIT'2001*, p. 219, Washington DC, June 2001.
19. W. Su, Z. Safar, M. Olfat, and K. J. R. Liu, Obtaining full-diversity space-frequency codes from space-time codes via mapping, *IEEE Trans. Signal Process.* (Special Issue on MIMO Wireless Communications), vol. 51, no. 11, pp. 2905–2916, November 2003.

20. W. Su, Z. Safar, and K. J. R. Liu, Full-rate full-diversity space-frequency codes with optimum coding advantage, IEEE Trans. Inform. Theory, vol. 51, no. 1, pp. 229–249, January 2005.

21. Y. Gong and K. B. Letaief, Space-frequency-time coded OFDM for broadband wireless communications, in *Proc. IEEE GLOBECOM*, San Antonio, USA, November 2001.

22. B. Lu, X. Wang, and K. R. Narayanan, LDPC-based space-time coded OFDM systems over correlated fading channels: Performance analysis and receiver design, *IEEE Trans. Commun.*, vol. 50, no. 1, pp. 74–88, January 2002.

23. A. F. Molisch, M. Z. Win, and J. H. Winters, Space-time-frequency (STF) coding for MIMO-OFDM systems, *IEEE Commun. Lett.*, vol. 6, no. 9, pp. 370–372, September 2002.

24. Z. Liu, Y. Xin, and G. Giannakis, Space-time-frequency coded OFDM over frequency selective fading channels, *IEEE Trans. Signal Proc.*, vol. 50, pp. 2465–2476, October 2002.

25. Z. Liu, Y. Xin, and G. B. Giannakis, Linear constellation precoding for OFDM with maximum multipath diversity and coding gains, *IEEE Trans. Commun.*, vol. 51, no. 3, pp. 416–427, March 2003.

26. W. Su, Z. Safar, and K. J. R. Liu, Towards maximum achievable diversity in space, time and frequency: Performance analysis and code design, *IEEE Trans. Wireless Commun.*, vol. 4, no. 4, pp. 1847–1857, July 2005.

27. W. Su, Z. Safar, and K. J. R. Liu, Space-time signal design for time-correlated Rayleigh fading channels, *Proc. IEEE ICC*, vol. 5, pp. 3175–3179, Anchorage, Alaska, May 2003.

28. C. Schlegel and D. J. Costello, Jr., Bandwidth efficient coding for fading channels: Code construction and performance analysis, *IEEE JSAC*, vol. 7, pp. 1356–1368, December 1989.

29. K. Boullé and J. C. Belfiore, Modulation schemes designed for the Rayleigh channel, *Proc. CISS'92*, pp. 288–293, Princeton University, NJ, March 1992.

30. X. Giraud, E. Boutillon, and J. C. Belfiore, Algebraic tools to build modulation schemes for fading channels, *IEEE Trans. Inform. Theory*, vol. 43, no. 2, pp. 938–952, May 1997.

31. J. Boutros and E. Viterbo, Signal space diversity: A power- and bandwidth-efficient diversity technique for the Rayleigh fading channel, *IEEE Trans. Inform. Theory*, vol. 44, no. 4, pp. 1453–1467, July 1998.

32. M. Damen, A. Chkeif, and J. Belfiore, Lattice code decoder for space-time codes, *IEEE Commun. Lett.*, vol. 4, no. 5, pp. 161–163, 2000.

33. Z. Safar, W. Su, and K. J. R. Liu, A fast sphere decoding algorithm for space-frequency codes, *EURASIP J. App. Signal Process.*, vol. 2006, pp. 1–14, 2006.

34. O. Y. Takeshita and D. J. Constello, Jr., New classes of algebraic interleavers for turbo-codes, in *Proc. ISIT*, p. 419, 1998.

35. H. S. Wang and N. Moayeri, Finite-state Markov channel – a useful model for radio communication channels, *IEEE Trans. Veh. Tech.*, vol. 44, no. 1, pp. 163–171, 1996.

36. G. Stuber, *Principles of Mobile Communication*, Kluwer Academic Publishers, Norwell, MA, 2001.

37. R. A. Horn and C. R. Johnson, *Topics in Matrix Analysis*, Cambridge Univ. Press, Cambridge, UK, 1991.

Part II

Performance Analysis

5

Performance Analysis of IEEE 802.16 Fixed Broadband Wireless Access Systems

R. Jayaparvathy and McNeil Ivan

CONTENTS

5.1 Introduction

Fixed broadband wireless access (FBWA) system is defined by the IEEE 802.16 standard. FBWA provides network access to buildings through exterior antennas communicating with central radio base stations (BSs). The IEEE standard 802.16-2004 specifies the air interface for FBWA systems supporting multimedia services [1]. The MAC supports a point-to-multipoint (PMP) architecture with the optional mesh topology. It is structured to support multiple physical layer (PHY) specifications each suited to a particular operational environment. For operating frequencies of 10–66 GHz, the wireless MAN-SC based on single carrier modulation is specified. For frequencies below 11 GHz, where the propagation without direct line of sight must be accommodated, the wireless MAN-OFDM (orthogonal frequency division multiplexing), the wireless

MAN-OFDMA (orthogonal frequency division multiplexed access), and the wireless MAN-SC (using single carrier modulation) are employed [2,3].

The MAC of the IEEE 802.16 has three sublayers. The service-specific convergence sublayer (CS) provides mapping of external data, received through the CS service access point (SAP), into MAC service data units (SDUs) received by the MAC common part sublayer (CPS) through the MAC SAP. This includes classifying external network SDUs and associating them to the proper MAC service flows and connection identifier (CID). It may also include functions such as payload header suppression. Multiple CS specifications are provided for interfacing with various protocols. The MAC CPS provides the core MAC functionality of system access, bandwidth allocation, connection establishment, and maintenance. It receives data from the various CSs, through the MAC SAP and classifies to particular MAC connections. The IEEE 802.16 standard defines the QoS signaling framework and various types of service flows, but the actual QoS mechanisms such as packet scheduling and admission control algorithms for these service flows are unspecified in the standard.

To satisfy the QoS requirements for real-time multimedia applications in IEEE 802.16, admission control and proper scheduling are needed. Admission control maintains an appropriate number of connections in the network and determines whether a new call should be admitted, with the decision depending on whether the network can support the QoS of the new call.

The IEEE 802.16 standard defines four types of service flows each with different QoS requirements, namely the unsolicited grant service (UGS), real-time polling service (rtPS), nonreal-time polling service (nrtPS), and best effort (BE). Bandwidth allocation is done through the request/grant mechanism. Bandwidth request is always made per connection and grants are either per connection (GPC) or per SS (GPSS) [3].

Very few analytical models have been presented in literature to analyze the performance of IEEE 802.16 FBWA systems. Lai and Lin [4] presented a model for call admission control in QoS capable networks using a multidimensional Markov chain approach and analyzed three types of buffer models (i.e., no buffer, buffer without preemption, and buffer with preemption). Ahmed et al. [5] presented the multimedia performance of IEEE 802.16 using two types of traffic, namely Ethernet packet traffic and constant bit-rate traffic, which were assumed to follow Poisson distribution. A complete survey of wireless MAC protocols, such as general network concepts, wireless MAC issues, performance metrics, classification of MAC protocols and comparison, were studied in Ref. 6. An architecture to support QoS mechanisms in IEEE 802.16 standard was developed, and the weighted fair queuing (WFQ) algorithm was used for evaluating the performance by Safi in Ref. 7. The architecture includes shaper and policer, four different queues, uplink service flow data base for both SS and BS in addition to grant allocator, request generator, and upstream generator for SS and downlink service flow data base, polling manager, uplink scheduler, downlink scheduler, and downstream generator for BS, respectively.

A pinwheel approach for the real time scheduling problem was studied by Feinberg and Curry [8] and Holte et al. [9]. Performance evaluation of IEEE 802.16 for broadband wireless access MAC layer that supports a range of physical layer technologies was presented by Ramachandran et al. in Ref. 10. An efficient uplink scheduling algorithm based on voice activity for VoIP services in IEEE 802.16d/e systems was presented by Lee et al. in Ref. 11. The authors have shown that the algorithm provides higher throughput performance and increase in number of users supported. Performance evaluation of various scheduling schemes was presented by Jayaparvathy et al. in Ref. 12. The authors proposed a scheduler based on the transmission opportunity of an SS considering three different queues for three types of services (rtPS, nrtPS, BE) assuming the bandwidth is already allocated to UGS services and have shown that the TXOP-based scheduler provides improved performance.

In this chapter, we propose a combined call admission control and scheduling scheme for the IEEE 802.16 system considering the bandwidth constraint for GPSS mode of operation. We develop an analytical model to evaluate the throughput performance of the system when the above scheme is implemented. Analysis of the IEEE 802.16 MAC protocol is done by varying the bandwidth of all types of services. We compare the performance of this algorithm with first-in first-out (FIFO) and transmission opportunity (TXOP) based scheduling schemes, and show that the proposed scheme provides improved performance at higher loads. We validate our analytical model through simulations.

The rest of the chapter is organized as follows. In Section 5.2, we present the QoS features of IEEE 802.16. The system model considered for performance analysis is presented in Section 5.3. The analytical model is presented in Section 5.4. We present the numerical results in Section 5.5. Section 5.6 presents the conclusions and further work.

5.2   QoS Features of IEEE 802.16

The IEEE 802.16 standard for FBWA systems supports the metropolitan area network architecture. It assumes a PMP topology with a BS and several SSs. BS controls and manages the entire system and SSs perform as interface between end users and the BS. The downlink channel on which data flow is directed from BS to SSs uses TDM scheme and the uplink channel uses TDMA scheme. The IEEE 802.16 standard defines four types of service flows, each with different QoS requirements [1] as given below.

(i) *Unsolicited grant service*: The UGS is designed to support real-time service flows that generate fixed-size data packets on a periodic basis, such as T1, E1, and voice over IP without silence suppression.

This service receives fixed size unsolicited data grants (transmission opportunities) on a periodic basis. Therefore, it eliminates the overhead and latency of requiring the SS to send requests for data grant.

(ii) *Real-time polling service*: The rtPS is designed to support real-time service flows that generate variable size data packets on a periodic basis, such as MPEG video. This service offers periodic unicast request opportunities, which meet the flow's real-time needs.

(iii) *Nonreal-time polling service*: This service is introduced for nonrealtime flows that require variable size data grants on a regular basis, such as high bandwidth FTP. This service offers unicast polls on a periodic basis, but using more spaced intervals than rtPS. This ensures that the flow receives request opportunities even during network congestion.

(iv) *Best effort service*: This service is for BE traffic, such as HTTP. There is no QoS guarantee. The IEEE 802.16 standard defines several ways for SSs to request bandwidth, combining the determinism of unicast polling with the responsiveness of contention-based requests and the efficiency of unsolicited bandwidth.

The BS is allowed to allocate bandwidth in the following two modes:

(i) Grant per connection (GPC), in which bandwidth is assigned to each connection, and

(ii) Grant per subscriber station (GPSS), in which an SS requests for transmission opportunities for all of its connections and redistributes the bandwidth among them.

The latter is more suitable when there exist many connections per terminal and it is mandatory for systems using the 10–66 GHz PHY specification [1]. With the grant per SS (GPSS) class, SSs are granted the bandwidth aggregated into a single grant to the SS itself. The GPSS SS needs to be more intelligent in its handling of QoS. It will typically use the bandwidth for the connection that requested it, but need not. For instance, if the QoS situation at the SS has changed since the last request, the SS has the option of sending the higher QoS data along with a request to replace this bandwidth stolen from a lower QoS connection. Therefore, there should be a proper admission control scheme such that the entire system throughput is increased and the channel should be properly utilized. Very few analytical models have been proposed for the systems working in the GPSS mode of operation.

5.3 System Model

Consider a system with one BS and five subscriber stations as shown in Figure 5.1. QoS architecture for the BS and the SS are considered as

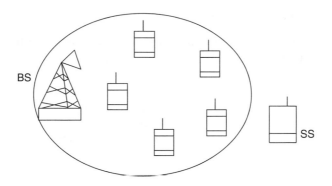

FIGURE 5.1
System model.

in Ref. 7. It is of interest to develop an analytical model to compute the throughput performance of the system. We consider four types of services, namely the rtPS, nrtPS, BE, and UGS. Bandwidth is assumed to be reserved for UGS.

The following assumptions are made for performance analysis of the IEEE 802.16 system.

- Call arrivals of rtPS, nrtPS, and BE traffic are assumed to be independent and generally distributed.
- The service times of all the three types of traffic are exponentially distributed.
- Three different G/M/1 infinite size queues with the different waiting time thresholds [2] for the different traffic types are assumed. rtPS service is considered to be delay intolerant.
- Upstream and downstream traffic are assumed to be similar.
- Self-correcting nature of the request/grant protocol requires that SSs shall periodically use aggregate bandwidth request.
- Aggregate bandwidth request interval is 40 ms as mentioned in Ref. 7 and incremental request is through piggybacking.

5.4   Performance Analysis

Consider the system shown in Figure 5.1. To obtain the throughput performance of the system it is necessary to model the queue. Since call arrivals follow a general distribution (Pareto) and service times are assumed to be exponentially distributed, the system is modeled as a G/M/1 queue. The G/M/1 queuing model is explained in detail in the following subsection.

5.4.1   G/M/1 Queuing Model

The service time distribution of calls is assumed to be exponential with mean $1/\mu$ and that the interarrival time is a random variable, having an arbitrary (general) distribution with mean $1/\lambda$. The queue is serviced by a single server and has an infinite waiting room. Generally, Pareto distribution is considered for the interarrival time as the calls coming into the system are self-similar and independent. We denote the density function of interarrival time by $A(t)$ and its probability density function (p.d.f), when it exists by $a(t)$.

For a Pareto distribution,

$$A(t) = 1 - \left(\frac{x_m}{t}\right)^k \quad k > 0 \tag{5.1}$$

where x_m is the minimum value of t.

Let $X(t)$ be the number of customers in the system at time t. $\{X(t), t \geq 0\}$ is a continuous time stochastic process with state space $\{0, 1, 2, ...\}$. Knowing the current state of $X(t)$ does not provide enough information about the time until the next arrival (unless the interarrival times are exponentially distributed), and hence the future cannot be predicted solely based on $X(t)$. The G/M/1 queue is stable if $\rho < 1$, where $\rho = \lambda/\mu$ is the traffic intensity as given in Ref. 13. The queue, is assumed to be stable for the remaining part of the analysis. To analyze the queue, it is necessary to find the limiting probability of the number of customers in the system that necessitates the functional equation.

5.4.2   Functional Equation

A key functional equation arises in the study of the limiting behavior of a G/M/1 queue. The key functional equation of a G/M/1 queue is as given below:

$$z = A^*(\mu(1 - z)) \tag{5.2}$$

$z = 1$ is always a solution to the key functional equation. There are other solutions to this equation in the interval $(0, 1)$. If $\rho < 1$, there is a unique solution $\alpha \in (0, 1)$ to the key functional equation [14]. It is not always possible to solve the key functional equation of the G/M/1 queue analytically. In such cases, the solution can be obtained numerically by using the following recursive computation:

$$z_0 = 0, \quad z_{n+1} = A^*(\mu(1 - z_n)) \tag{5.3}$$

Then, if $\rho < 1$,

$$\lim_{n \to \infty} Z_n = \alpha \tag{5.4}$$

where α is the unique solution in $(0, 1)$ to the key functional equation.

Since the process $\{X(t), t \geq 0\}$ jumps by $+1$ or -1, it follows that the arrival time distribution is the same as the departure time distribution,

$$\Pi_j = \Pi_j^* \quad j \geq 0 \tag{5.5}$$

In a stable $G/M/1$ queue, the limiting distribution of the number of customers as seen by an arrival is given by

$$\Pi_j^* = (1 - \alpha)^* \alpha^j \quad j \geq 0$$

where α is the unique solution in $(0, 1)$ to the key functional equation. Since the arrival in a $G/M/1$ queue is not Poisson, the limiting distribution is not equal to the arrival distribution. In a stable $G/M/1$ queue, the limiting distribution of the number of customers in the system is given by

$$p_0 - 1 - \rho \tag{5.6}$$

$$p_j = \rho^* \Pi_{j-1}^* \quad j \geq 1 \tag{5.7}$$

5.4.3   Power-Tail Distributions

Internet traffic is not Poisson in nature. It is found to follow a heavy-tailed distribution. Pareto and Power are a subclass of heavy-tailed distribution. A heavy-tailed distribution is one whose tail probability decays more slowly than any exponential function. Researchers have shown that many distributions associated with Internet traffic are heavy-tailed [15]. Power-tail distributions have been found appropriate to capture the burstiness of network traffic over multiple time scales. Pareto is a well-known power-tail distribution. A random variable X has a power-tail distribution if its complementary cumulative distribution function (ccdf) satisfies

$$\overline{F}(t) = \Pr\{X > t\} \sim ct^{-\alpha} \quad \text{as } t \to \infty \tag{5.8}$$

where α and c are positive constants.

A well-known power-tail distribution is the (translated) Pareto distribution for which

$$\overline{F}(t) = \Pr\{X > t\} = \frac{1}{(1 + at)^\alpha} \quad \text{for } t \geq 0 \text{ and } a > 0 \tag{5.9}$$

The Pareto distribution provides parsimonious modeling since it depends on only two parameters.

Unfortunately, power-tail distributions do not lend themselves to easy queuing analysis since their Laplace transforms are not explicit. This explains why, so far, most of the queuing results involving power-tail distributions have only been obtained in asymptotic regimes. These asymptotic results have the merit of providing some insight into the relation between the power-tail distribution parameters and the queuing performance measures [16].

Analyzing queues with heavy-tailed distributions is difficult because these distributions do not have closed-form, analytic Laplace transforms. Thus, analytical methods do not work. One way around this is to approximate a heavy-tailed distribution with a phase-type distribution. We apply the approximation method known as the transform approximation method (TAM) developed for power-tailed distributions as in Ref. 16. To obtain more quantitative results, several contributions have been recently made fitting hyper-exponential distributions. A methodology for approximating power-tail distributions into hyper-exponential distributions is obtained as in Ref. 17. This algorithm is slightly modified according to the arrival distributions.

A new methodology for fitting hyper-exponential distributions to power-tail distributions is obtained following Ref. 16. This new approach exhibits several advantages. First, the approximation can be made arbitrarily close to the exact distribution and bounds on the approximation error are easily obtained. Second, the fitted hyper-exponential distribution depends only on a few parameters that are explicitly related to the parameters of the power-tail distribution. Third, only a small number of exponentials are required to obtain an accurate approximation over multiple time scales, e.g., a dozen of exponentials for five time scales. Once equipped with a fitted hyper-exponential distribution, an integrated framework for analyzing queuing systems with power-tail distributions is obtained as in Ref. 16.

5.4.4 The Fitting Algorithm

In this section, we explain the fitting algorithm used in Refs. 16 and 17. Our algorithm proceeds in two stages. The first and most significant stage focuses on fitting a mixture of exponentials to the behavior of the tail of the power-tail distribution. The second stage provides a fitting for small values of t and ensures that the mixture of exponentials is indeed a probability distribution. As an example, we consider the case of the Pareto distribution defined in Equation 5.8.

Consider the function $R(t) = ct^{-\alpha}$. We want to derive an expression for a mixture of exponentials that can capture the behavior of $R(t)$ from some value of t and over an arbitrary large number of timescales. Our starting point relies on the fact that $ct^{-\alpha}$ is the Laplace transform of the function $a(s) = cs^{\alpha-1}/\Gamma(\alpha)$, where $\Gamma(\cdot)$ is the Gamma function. We can, therefore, express $R(t)$ in the following way

$$R(t) = c \int_{0}^{\infty} \frac{s^{\alpha-1} e^{-st}}{\Gamma(\alpha)} \, ds \qquad (5.10)$$

We let $c = 1$ since it is merely a constant of proportionality. The integral appearing in the right-hand side of Equation 5.10 can be approximated by a Riemann sum. However, according to Tauberian theorems [18] the behavior of $R(t)$ for large values of t is closely related to the behavior of $r(s)$ near $s = 0$. The choice of a fixed grid would not be wise. It would put too much emphasis on large values of s corresponding to high frequencies and not enough on small values of s corresponding to low frequencies. We perform the following change of variables from s to u, $s = B^{-u}$, where $B > 1$ is a parameter that controls the accuracy of the approximation. We note that choosing a fixed grid for the variable u is equivalent to choosing a logarithmic grid for s. After the change of variables, Equation 5.10 can be rewritten as

$$R(t) = \frac{\log B}{\Gamma(\alpha)} \int_{-\infty}^{\infty} B^{-\alpha u} \exp(-tB^{-u}) \, du$$

$$= \frac{\log B}{\Gamma(\alpha)} \sum_{n=-\infty}^{\infty} \int_{n-1/2}^{n+1/2} B^{-\alpha u} \exp(-tB^{-u}) \, du \qquad (5.11)$$

Equation 5.11 can be approximated by a Riemann sum if we replace each integrand with its mid-span value. It turns out, however, that a better approximation can be obtained if only the exponent portion of the integrand is replaced with its mid-span value. We have then

$$R(t) \approx \frac{\log B}{\Gamma(\alpha)} \sum_{n=-\infty}^{\infty} \exp(-tB^{-n}) \int_{n-1/2}^{n+1/2} B^{-\alpha u} \, du$$

$$= \frac{B^{\alpha/2} - B^{-\alpha/2}}{\Gamma(\alpha+1)} \sum_{n=-\infty}^{\infty} B^{-\alpha n} \exp(-tB^{-n}) \equiv R_1(t) \qquad (5.12)$$

with $B \to 1$, the approximation $R_1(t)$ can be made arbitrarily close to $R(t)$. The last step of the algorithm is to truncate the infinite sum $R_1(t)$ and approximate it by a finite sum $R_2(t)$, where

$$R_2(t) = \frac{B^{\alpha/2} - B^{-\alpha/2}}{\Gamma(\alpha+1)} \sum_{n=M}^{N} B^{-\alpha n} \exp(-tB^{-n}) \qquad (5.13)$$

The idea behind this truncation is the following. On the one hand, values of n below M correspond to high frequencies that have almost no effect on the long-term behavior of $R(t)$. On the other hand, values of n larger than N correspond to very low frequencies (or very large values of t) falling beyond the scope of interest. We note that the approximation $R_2(t)$ is very parsimonious since it depends on only four parameters α, B, M, and N. Figure 5.2 shows the variation of $R_2(t)$ as a function of t.

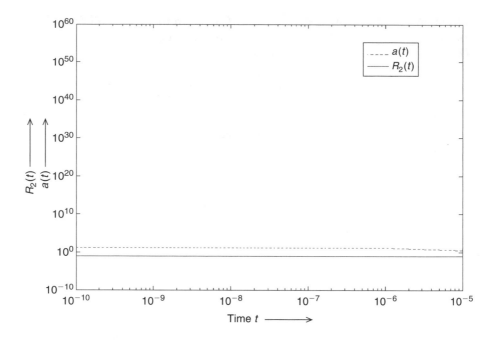

FIGURE 5.2
Approximation plots.

We denote the arrival distribution of the various types of traffic by $a(t)$. From Figure 5.2, it can be seen that $a(t)$ directly correlates to $R_2(t)$. The Laplace transform of the arrival distribution can be obtained from the approximated functions. The values of constants M, N, and B are chosen as in Ref. 17, respectively, to obtain the approximated arrival distribution of the three classes of traffic as follows,

$$\text{rtPS}: a(t) = 1.79 * \exp(-0.167 * \lambda_1 * t) * [9.11 - (0.754 * 10^6 * t)] \quad (5.14)$$

$$\text{nrtPS}: a(t) = 1.79 * \exp(-0.167 * \lambda_2 * t) * [6.48 - (0.155 * 10^6 * t)] \quad (5.15)$$

$$\text{B.E}: a(t) = 1.79 * \exp(-0.167 * \lambda_3 * t) * [5.853 - (0.09165 * 10^6 * t)] \quad (5.16)$$

The system contains three different queues for each of the three types of traffic namely rtPS, nrtPS, and BE. The system is analyzed as three different G/M/1 queues and the probability of number of users (p_j) is found for all the three queues separately. The value of α for all the three queues is found using Equations 5.14 through 5.16. The states of the system are formulated considering the bandwidth of all traffic types and the total channel capacity allocated to a particular SS. (n_1, n_2, n_3) is denoted as the state space, where n_1, n_2, and n_3 are the numbers of type-1, type-2, and type-3 calls in the system, respectively. If the residual bandwidth is not sufficient for an incoming call and the buffer is empty, this call gets queued in this free buffer. The above situation is denoted as a state with an underline mark, namely, $(\underline{n_1}, n_2, n_3), (n_1, \underline{n_2}, n_3)$, or $(n_1, n_2, \underline{n_3})$. State (n_1, n_2, n_3) means that $(n_1$-1) type-1 calls, n_2 type-2 calls, and n_3 type-3 calls are in service, and one type-1 call is waiting in the buffer. We define an algorithm to construct all states. The algorithm for state-space construction is as follows:

```
for {n3 = 0 to [C/B3]
    for {n2 = 0 to [(C-n2B2)/B2]
        for {n1 = 0 to [(C-n2B2-n3B3)/B1]
            Create state (n1, n2, n3) }}}
```

The call arrivals in these three queues are independent of each other. We use this property to find the probability that the system is in a particular state. The probability that the system exists in a state is determined by the limiting probability of the three queues. Probability that the state is $(0, 0, 0) =$ (probability that there are 0 calls in the first queue) * (probability that there are 0 calls in the second queue) * (probability that there are 0 calls in the third queue).

Similarly, the probability of all the stable states (states in which there are no calls waiting to be serviced) can be found. The states that have enough residual bandwidth for an incoming call of a particular class are added up to determine the probability that there is enough residual bandwidth for that class of traffic. For example, with $B_1 = 12$ Mbps, $B_2 = 8$ Mbps, $B_3 = 8$ Mbps, and $C = 30$ Mbps the state (0,0,1) has enough residual bandwidth for an incoming rtPS, but state (2,0,0) does not have enough residual bandwidth for an incoming rtPS call to be serviced.

Pr (bandwidth is free for an incoming class k traffic) = Σ (probability of states that have enough residual bandwidth for that class of traffic).

5.4.5 Throughput of a Class of Traffic

The throughput of a particular class of traffic depends upon three factors as follows:

1. The number of calls of a particular type of service that occupies the given channel capacity. For example, if $B_1 = 2$ Mbps and $C = 30$ Mbps, then the throughput of the service will be very high as the residual bandwidth will be very high and more calls of that type can be serviced.

2. The bandwidth occupied by the other class of traffic. If the bandwidth occupied by type-1 and type-2 classes of traffic is very high then the residual bandwidth for type-1 call will be low, and the type-1 call has to wait and has a high probability of being lost thereby decreasing the throughput.

3. The usage of residual bandwidth by a class of traffic. If there is residual bandwidth in the system, then the probability with which that residual bandwidth will be used for the incoming traffic should be taken into consideration.

The throughput of the three classes of service are given by the following equations,

$$rtPS = B_1/C * [n_1 + (R_1/B_1 * Pr(\text{bandwidth is free for rtPS})) - (B_2 * B_3/C^2)^{0.5}]$$
(5.17)

$$nrtPS = B_2/C * [n_2 + (R_2/B_2 * Pr(\text{bandwidth is free for nrtPS}))$$
$$- (B_1 * B_3/C^2)^{0.5}]$$
(5.18)

$$BE = B_3/C * [n_3 + (R_3/B_3 * Pr(\text{bandwidth is free for BE}))$$
$$- (B_2 * B_1/C^2)^{0.5}]$$
(5.19)

where

B_1, B_2, B_3—Bandwidths of rtPS, nrtPS, and BE, respectively
C—Total channel capacity allocated to the SS
n_1, n_2, n_3—number of users occupied by the rtPS, nrtPS, and BE types
$n_1 = $ integer (C/B_1), $n_2 = $ integer (C/B_2), $n_3 = $ integer (C/B_3)
$R_1, R_2, R_3 = $ Residual bandwidth if the whole channel is occupied by rtPS, nrtPS, and BE
$R_1 = C - (n_1 * B_1), R_2 = C - (n_2 * B_2), R_3 = C - (n_3 * B_3)$

The average throughput of a class of traffic is given by

$$\text{Throughput}_k = \frac{\text{Number of calls of type } k \text{ that are successfully serviced}}{\text{Number of calls of type } k \text{ generated}}$$
(5.20)

where k denotes the class of traffic.

5.4.6 Simulation Model

Performance evaluation is performed under three different scenarios, as follows:

(i) Vary the traffic load of rtPS and observe the throughput of rtPS, nrtps, and BE services
(ii) Vary the traffic load of nrtPS and observe the throughput of rtPS, nrtPS, and BE services
(iii) Vary the traffic load of BE and observe the throughput of rtPS, nrtPS, and BE services

TABLE 5.1

Parameters Used in Simulation

Cell radius	1 km
Duplexing schemes	TDD
Ratio of uplink slot to downlink in TDD	50%
Number of slots per frame	5000
Number of subscriber stations	5
Downstream data transmission rate	20 Mbps
Aggregate upstream data transmission rate	30 Mbps
Initial backoff parameter	3 (window size = 8)
Maximum backoff parameter	10 (window size = 1024)

The parameters used in simulation are listed in Table 5.1.

Results are tabulated by maintaining the bandwidth allocated to UGS traffic and increasing rtPS, nrtPS, and BE traffic loads in steps from 2 to 14 Mbps.

5.5 Results and Discussion

In this section, we present the numerical results obtained from our analytical model with admission control and scheduling. We validate our analysis by comparing the results obtained with that of simulations. Figure 5.3 presents rtPS average throughput as a function of the traffic load of rtPS for the FIFO, TXOP-based allocation, and the proposed admission control and scheduling schemes. It can be observed from Figure 5.3 that the FIFO scheduling scheme provides lesser throughput as traffic load increases, the TXOP-based allocation provides consistently high throughput whereas with the proposed scheme as the traffic load of rtPS increases, the throughput decreases because of the increase in the overall traffic load in the system and as the bandwidth of the rtPS increases, the residual bandwidth for rtPS decreases resulting in decrease in throughput. When the traffic load of rtPS is very low, the calls can be serviced without waiting or being dropped, so the throughput is very high.

Figure 5.4 presents the nrtPS average throughput as a function of the traffic load of rtPS. From Figure 5.4, it can be observed that there is a decrease in nrtPS throughput at higher loads of rtPS. With the proposed scheme, as the traffic load of rtPS increases, the throughput of nrtPS decreases linearly because of the increase in the overall traffic load in the system.

Figure 5.5 presents BE average throughput as a function of the traffic load of rtPS. It can be observed from Figure 5.5 that, with FIFO scheme, a high throughput is obtained as the BE traffic is allocated resources similar to the rtPS and nrtPS traffic classes. With the TXOP-based allocation scheme, the throughput performance decreases with increase in traffic load. With the proposed scheme, as the traffic load of rtPS increases, the throughput of BE decreases because it increase the overall traffic load in the system and as the bandwidth of the rtPS increases, the residual bandwidth for rtPS also decreases resulting in decrease in throughput.

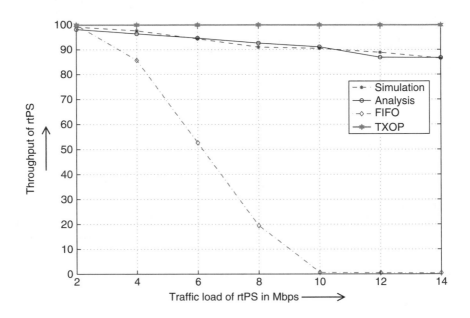

FIGURE 5.3
Throughput of rtPS versus traffic load of rtPS.

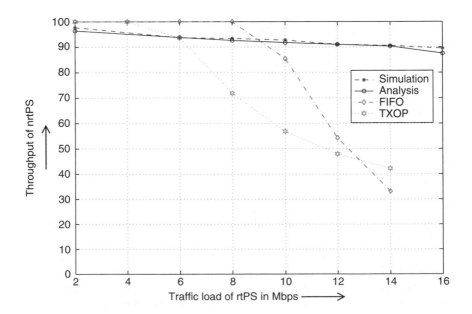

FIGURE 5.4
Throughput of nrtPS versus traffic load of rtPS.

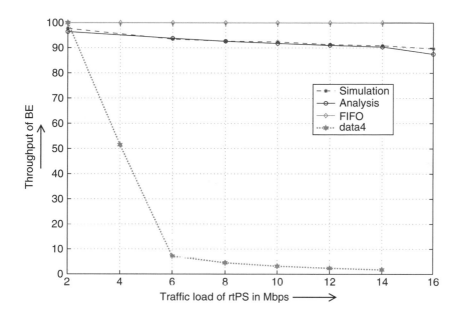

FIGURE 5.5
Throughput of BE versus traffic load of rtPS.

Figure 5.6 presents the rtPS average throughput as a function of the traffic load of nrtPS service. The throughput of the rtPS service decreases linearly as the traffic load of nrtPS service is increased because the overall traffic load into the system increases. As the bandwidth of rtPS is assumed to be 12 Mbps, there will not be enough bandwidth for the next rtPS call coming into the system and the call may have to wait or dropped, so the throughput of the rtPS decreases as the traffic load of nrtPS is increased. The TXOP-based allocation scheme provides improved throughput performance whereas the FIFO scheme provides reduced throughput with increase in nrtPS traffic load.

Figure 5.7 presents the nrtPS average throughput as a function of the traffic load of nrtPS. When the traffic load of the nrtPS is very low, there will be enough bandwidth for nrtPS to get serviced, but as the traffic load increases the residual bandwidth decreases and the throughput of nrtPS decreases linearly. The proposed scheme is better at higher traffic loads compared to the FIFO and the TXOP-based allocation schemes.

Figure 5.8 presents the BE average throughput as a function of traffic load of nrtPS. With the proposed scheme, when the traffic load of the nrtPS is very low, there will not be enough bandwidth for a BE call to get serviced as the priority is very high for nrtPS call, but as the traffic load increases the residual bandwidth for nrtPS decreases and the residual bandwidth for BE service is available for its service and so the throughput of BE has a very small slope in its graph (remains almost like a straight line). The performance of the FIFO

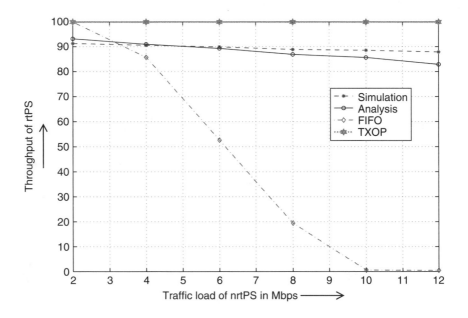

FIGURE 5.6
Throughput of rtPS versus traffic load of nrtPS.

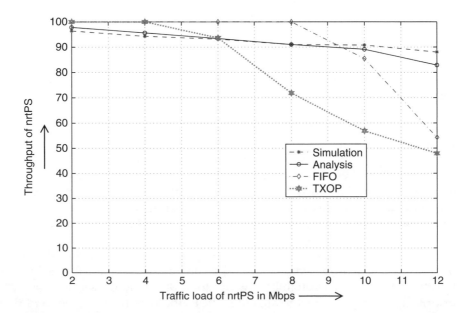

FIGURE 5.7
Throughput of nrtPS versus traffic load of nrtPS.

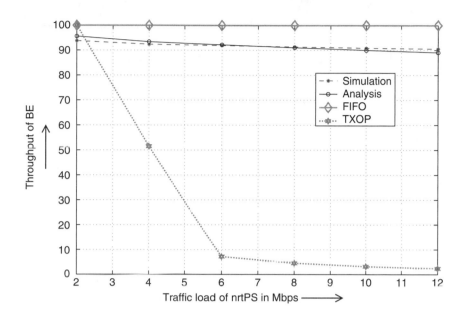

FIGURE 5.8
Throughput of BE versus traffic load of nrtPS.

scheme is better in this case whereas TXOP-based allocation scheme provides reduced throughput at higher loads.

Figure 5.9 presents the rtPS average throughput as a function of the traffic load of BE service. With the proposed scheme, the throughput of the rtPS service decreases linearly as the traffic load of nrtPS service is increased because the overall traffic load into the system increases. As the bandwidth of rtPS is assumed to be 12 Mbps, there will not be enough bandwidth for the next rtPS call coming into the system and the call may have to wait or get dropped. The increase in the overall traffic load into the system will decrease the throughput of rtPS call, so the throughput of the rtPS decreases as the traffic load of BE is increased. While the TXOP-based allocation scheme performs better, the FIFO scheme performs the worst at higher traffic loads.

Figure 5.10 presents the nrtPS average throughput as a function of the traffic load of BE service. With the proposed scheme as the traffic load of the system increases, the throughput of nrtPS traffic decreases. Moreover, nrtPS calls can get bandwidth only after the rtPS calls are serviced, so the throughput decreases; but as it has higher priority over BE traffic, the throughput does not decrease rapidly. Both TXOP- and FIFO-based schemes provide decreased throughput at higher loads in this case.

Figure 5.11 presents the BE average throughput as a function of the traffic load of BE. With the proposed scheme, when the traffic load of the BE is very low, there will be enough bandwidth for an nrtPS to get serviced, but BE traffic is the least priority service, so even for lower traffic loads the throughput of

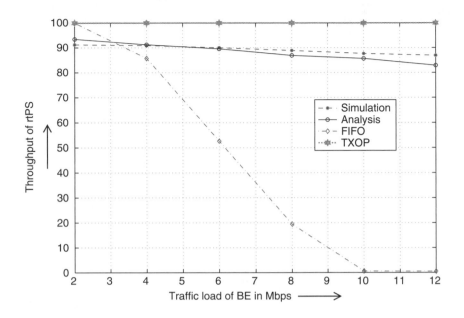

FIGURE 5.9
Throughput of rtPS versus traffic load of BE.

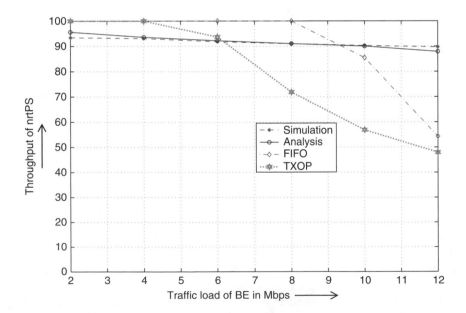

FIGURE 5.10
Throughput of nrtPS versus traffic load of BE.

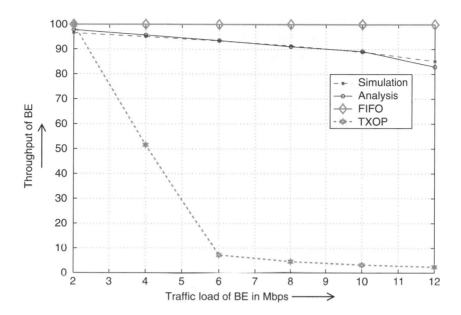

FIGURE 5.11
Throughput of BE versus traffic load of BE.

BE is lesser than rtPS, nrtPS services. As the traffic load increases the residual bandwidth decreases and the throughput of BE further decreases. The FIFO scheme performs best but the TXOP-based allocation scheme provides lesser throughput as traffic load is increased.

5.6 Conclusions and Future Work

In this chapter, we have proposed a call admission control and scheduling scheme for the IEEE 802.16 system considering the bandwidth constraint for GPSS mode of operation. Priority of the calls to support QoS in the IEEE 802.16 standard is also taken into account. We have developed an analytical model to evaluate the throughput and delay performance of the system when the above scheme is implemented. Analysis of the IEEE 802.16 MAC protocol is done by varying the bandwidth of all the types of services. We have compared the performance of this algorithm with FIFO- and TXOP-based schemes and show that our scheme provides improved performance at higher loads. Our analytical model has been validated through simulations. We show that the average system throughput of rtPS, nrtPS, and BE traffic have increased when a queue is employed for each of the traffic types. Comparison with FIFO and TXOP schedulers show that there is a greater performance improvement with our scheme.

Understanding IEEE 802.16 MAC protocol behavior for the various QoS mechanisms defined, different queuing algorithms and different types of services is important. Modeling the system will describe the complete behavior of the system. Further optimization can be done for determining priorities of the incoming call. Optimization techniques and evolutionary algorithms could be incorporated to efficiently utilize the bandwidth. Bandwidth reservation schemes can be employed to improve the throughput of the system. To study heavy-tailed distribution, more accurate models can be developed.

References

1. IEEE P802.16-Revd/D3—2004, *Draft Amendment to IEEE Standard for Local and Metropolitan Area Networks: Part 16: Air Interface for Fixed Access Wireless Systems— Medium Access Control Modifications and Additional Physical Layer Specifications for 2–11 GHz*, 2004.
2. IEEE 802.16d Standard, *IEEE Standard For Local and Metropolitan Area Networks: Media Access Control (MAC) Bridges*, 2004.
3. C. Eklund, R. B. Marks, K. L. Stanwood, and S. Wang, IEEE standard 802.16: a technical overview of the wireless MAN air interface for broadband wireless access, *IEEE Communication Magazine*, vol. 40, no. 6, pp. 98–107, Jun. 2002.
4. Y. C. Lai and Y. D. Lin, Fair admission control in QoS capable networks, *IEE Proceedings Communications*, vol. 152, no. 1, pp. 22–27, Feb. 2005.
5. A. Ahmad, C. Xin, F. He, and M. McKormic, *Multimedia Performance of IEEE 802.16 MAC*, Norfolk State University, Apr. 2005.
6. A. Chandra, V. Gummalla, and J. O. Limb, Wireless medium access control protocols, *IEEE Communications Surveys and Tutorials*, pp. 2–14, Second quarter, 2000.
7. E. Safi, *A Quality of Service Architecture for IEEE 802.16 Standard*, University of Toronto, 2004.
8. E. A. Feinberg and M. T. Curry, Generalized pinwheel problem, *Mathematical Methods of Operations Research*, vol. 62, pp. 99–122, 2005.
9. R. Holte, A. Mok, L. Rosier, I. Tulchinsky, and D. Varvel, The Pinwheel: A Real-Time Scheduling Problem, *22nd Hawaii International Conference of Systems Science*, pp. 693–702, Jun. 1989.
10. S. Ramachandran, C. W. Bostian, and S. F. Midkiff, Performance Evaluation of IEEE 802.16 for Broadband Wireless Access, *Proceedings of OPNETWORK*, 2002.
11. H. Lee, T. Kwon, and D.-H. Cho, An efficient uplink scheduling algorithm based on voice activity for VoIP services in IEEE 802.16 d/e systems, *IEEE Communication Letters*, vol. 9, no. 8, Aug. 2005.
12. R. Jayaparvathy, G. Sureshkumar, and P. Kankasabapathy, Performance evaluation of scheduling schemes for fixed broadband wireless access systems, *IEEE MICC ICON 2005*, Kaulalampur, Malaysia, 2005.
13. M. J. Fischer and T. B. Fowler, *Fractals, Heavy Tails and the Internet*, Sigma, Mitretek Systems, McLean, VA, Fall 2001.
14. J. Medhi, *Stochastic Processes*, John Wiley & Sons Inc., New York, July, 1994.

15. J. F. Shortle, M. J. Fischer, D. Gross, and D. M. B. Masi, Using the transform approximation method to analyze queues with heavy-tailed service, *Journal of Probability and Statistical Science*, vol. 1, no. 1, pp. 15–27, 2003.
16. B. Mandelbrot, A fast fractional Gaussian noise generator, *Water Resources Res.*, vol. 7, no. 3, pp. 543–553, 1971.
17. D. Starobinski and M. Sidi, Modeling and analysis of power-tail distributions, *Queuing Systems*, no. 36, pp. 243–267, 2000.
18. V. G. Kulkarni, *Modeling, Analysis, Design and Control of Stochastic Systems*, Springer, New York, 1999.
19. W. Feller, *An Introduction to Probability Theory and its Applications*, 2nd edition, vol. II, Wiley, New York, 1971.

6

System Performance Analysis for the Mesh Mode of IEEE 802.16

Min Cao and Qian Zhang

CONTENTS

6.1 Introduction

The rapid increase of user demands for faster connection to the Internet service has spurred broadband access network technologies advancement over recent years. While the backbone networks are matured and reliable with large bandwidth, the "last mile" has remained the bottleneck to enable broadband applications [1]. In the past few years, IEEE 802.11-based Wi-Fi networks have been widely deployed in hotspots, offices, campus, and airports to provide ubiquitous wireless coverage. However, this standard is handicapped in its short transmission range, bandwidth, quality of service (QoS), and security. The recently finalized IEEE 802.16 standard [2], also commonly known as WiMAX, is emerging as a promising broadband wireless technology to finally resolve the "last mile" problem in conjunction with IEEE 802.11.

IEEE 802.16 targets at providing last-mile fixed wireless broadband access in the metropolitan area network (MAN) with performance comparable to traditional cable, DSL, or T1 networks [3,15–18]. It operates at 10–66 GHz for line-of-sight (LOS), and 2–11 GHz for non-LOS connection, with a typi-cal channel bandwidth of 25 or 28 MHz. In the physical layer, the standard employs orthogonal frequency division multiplexing (OFDM), and supports adaptive modulation and coding depending on the channel conditions. It provides a high data rate of up to 134.4 Mbits/s, and a coverage up to 5 mi, as compared to several hundred feet in IEEE 802.11. The WiMAX deploy-ment not only serves the residential and enterprise users, but it can also be deployed as a backhaul for Wi-Fi hotspots or 3G cellular towers.

An IEEE 802.16 network consists of base station (BS) and subscriber station (SS). The BS is a node with wired connection and serves as a gateway for the SSs to the external network. The SSs are typically access point that aggregates traffic from end users in a certain geographical area. IEEE 802.16 supports two modes of operation: point-to-multipoint (PMP) mode and mesh mode. In PMP, each SS directly communicates with the BS through a single-hop link, which requires all SSs to be within clear LOS transmission range of the BS, as shown in Figure 6.1 (from Nokia White Paper [4]). In addition to the PMP mode, the MAC layer in IEEE 802.16a [5], which was integrated into IEEE 802.16-2004 [2], defines the control mechanisms and management mes-sages to establish connections in mesh network architectures. In the mesh mode, the SSs can communicate with the mesh BS and with each other through multihop routes via other SSs, as shown in Figure 6.2. The mesh topology not only extends the network coverage and reduces deployment cost in non-LOS environments, but also enables fast and flexible network configuration. Furthermore, the existence of multiple routes provides high

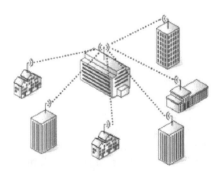

FIGURE 6.1
An example of an IEEE 802.16 PMP network.

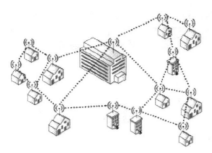

FIGURE 6.2
An example of an IEEE 802.16 mesh network.

network reliability and availability when node or link failures occur, or when channel conditions are poor. And with an intelligent routing protocol, the traffic can be routed to avoid the congested area. Owing to these advantages, WiMAX mesh networks provide a cost-effective solution for high speed "last mile" and backhaul applications in both metropolitan and rural areas.

In this chapter, we focus on the performance analysis of the IEEE 802.16 mesh mode, especially when the transmission is distributed coordinated. First, we give a brief overview of the IEEE 802.16 mesh mode in Section 6.2. In Section 6.3, we develop a model for assessing the performance of the distributed scheduler. We present the simulation methodology and results in Section 6.4 and compare the results with those in Section 6.3. Finally, Section 6.5 contains the conclusions.

6.2 Overview of IEEE 802.16 Mesh Mode

The IEEE 802.16 mesh mode uses time division multiple access (TDMA) for channel access among the mesh BS and SS nodes, where a radio channel is divided into frames. Each frame is further divided into time slots that can be assigned to the BS or different SS nodes. Figure 6.3 shows the frame structure in the mesh mode. A frame consists of a control subframe and a data subframe. Each frame is further divided into 256 minislots for transmission of user data and control messages. In the control subframe, transmission opportunities, which typically consist of multiple minislots, are used to carry signaling messages for network configuration and scheduling of data subframe minislot allocation. The transmission opportunity and minislot are the basic unit for resource allocation in the control and data subframes, respectively. There are two types of control subframes: network control subframe and scheduling control subframe. A network control subframe follows after every scheduling frames scheduling control subframes, where scheduling frames is a configurable network parameter.

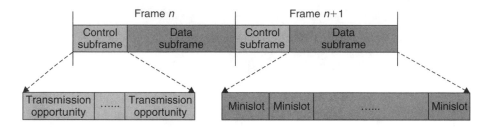

FIGURE 6.3
The IEEE 802.16 mesh frame structure.

In the network control subframes, mesh network configuration (MSH-NCFG) and mesh network entry (MSH-NENT) messages are transmitted for creation and maintenance of the network configuration. A scheduling tree rooted at the mesh BS is established for the routing path between each SS and the mesh BS. Active nodes within the mesh network periodically advertise MSH-NCFG messages, which contain a network descriptor that includes the network configuration information. A new node that wishes to join the mesh network scans for active networks by listening to MSH-NCFG messages. Upon receiving the MSH-NCFG message, the new node establishes synchronization with the mesh network. From among all the possible neighbor nodes that advertise MSH-NCFG, the new node selects one as its sponsor node. Then the new node sends a MSH-NENT message with registration information to the mesh BS through the sponsor node. Upon receipt of the registration message, the mesh BS adds the new node as the child of the sponsor node in the scheduling tree, and then broadcasts the updated network configuration to all SSs.

In the IEEE 802.16 mesh mode, both centralized scheduling and distributed scheduling are supported. Mesh centralized schedule (MSH-CSCH) and mesh distributed schedule (MSH-DSCH) messages are exchanged in the scheduling control subframe to assign the data minislots to different stations. The number of transmission opportunities for MSH-CSCH and MSH-DSCH in each scheduling control subframe are network parameters that can be configured. Centralized scheduling is mainly used to transfer data between the mesh BS and the SSs, while distributed scheduling targets data delivery between any two stations (BS or SS) in the same WiMAX mesh network. In the standard, the data subframe is partitioned into two parts for the two scheduling mechanisms, respectively.

The centralized scheduling handles both the uplink, where the traffic goes from the SSs to the mesh BS, and the downlink, where the traffic goes from the mesh BS to the SSs. In the mesh mode, time division duplex (TDD) is used to share the channel between the uplink and the downlink. In centralized scheduling, the mesh BS acts as the centralized scheduler and determines the allocation of the minislots dedicated to centralized scheduling among all the stations. The time period for centralized scheduling is called scheduling period, which is typically a couple of frames in length. There are two stages in each scheduling period. In the first stage, the SSs send bandwidth requests using the *MSH-CSCH:Request* message to their sponsor nodes, which are routed to the mesh BS along the scheduling tree. Each SS not only send its own bandwidth request, but also relays that of all its descendants in the scheduling tree. The SSs transmit *MSH-CSCH:Request* messages in such an order that the sponsor nodes always transmit after all their children. In this way, the mesh BS collects bandwidth requests from all the SSs. In the second stage, the mesh BS calculates and distributes the schedule by broadcasting the *MSH-CSCH:Grant* message, which is propagated to all the SSs along the scheduling tree. Because all the control and data packets need to go through the BS, the scheduling procedure is simple, but the connection setup delay is long. Hence, the centralized scheduling is not suitable for occasional traffic needs [6].

In distributed scheduling, the nodes are organized in an *ad hoc* fashion, and all nodes are peers and can act as routers to relay packets for their neighbors. Every node competes for channel access using a pseudorandom election algorithm based on the scheduling information of the two-hop neighbors, and data subframes are allocated through a request-grant-confirm three-way handshaking procedure. Hence, it is more flexible and efficient on connection setup and data transmission.

In distributed scheduling, the channel access in the control subframe is coordinated in a distributed manner among two-hop neighbors, and the data subframe slot allocation is performed through the control message exchange, so that there is no contention in the data subframe. Each node competes for transmission opportunities in the control subframe based on its neighbors' scheduling information such that in a two-hop neighborhood, only one node can broadcast its control message at any time. Once a node wins the control channel, a range of consecutive transmission opportunities are allocated to this node, which is called an eligible interval and the node can transmit in any slot in the interval. Every node needs to determine its next eligible interval during the current one. A pseudorandom function-based distributed election algorithm defined in the standard is used to decide whether the node wins a candidate transmission opportunity. If it wins, the reservation information is broadcast to the neighbors, otherwise the next slot is selected as a candidate and the procedure repeats until the node wins.

The control message used for distributed scheduling is mesh distributed scheduling (MSH-DSCH). The MSH-DSCH message of each node contains the schedule and data subframe allocation information of its one-hop neighbors as well as its own. By broadcasting the MSH-DSCH messages, each node can have the scheduling information of its two-hop neighbors. The next transmission time for every node can be computed based on such information. In the MSH-DSCH message, two parameters are included for control channel scheduling—Next_Xmt_Mx and Xmt_Holdoff_Exponent. The first parameter indicates the sequence number of the first slot in the eligible interval and the eligible length is $L = 2^{\text{Xmt\_Holdoff\_Exponent}}$ transmission opportunities. In order that each node can have the chance to access the control subframe, the MAC protocol requires every node to hold off some time before selecting the next

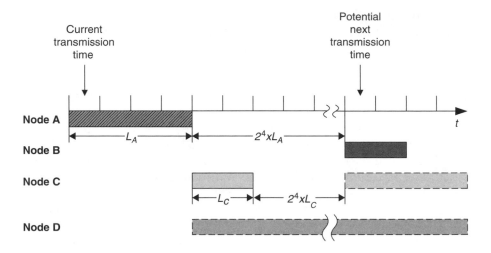

FIGURE 6.4
The IEEE 802.16 schedule control subframe contention.

transmission slot. In the standard, the holdoff time is defined to be 16 times of the eligible interval. The holdoff exponent value can decide a node channel contention frequency and affect all the nodes in two-hop neighborhood. In Sections 6.3 and 6.4, we can see that the holdoff exponent affects the scheduler performance significantly.

The control subframe competition procedure can be illustrated with Figure 6.4. During the current eligible interval (with length L_A), node A needs to determine its next transmission time. It will first compete for the first transmission opportunity that is holdoff time slots (which is $16L_A$ in length) after the current eligible interval. This transmission opportunity is called the temporary transmission slot. In the mesh mode, there are three types of competing nodes: (1) nodes whose the eligibility interval includes the temporary transmission slot (Node B); (2) nodes whose earliest future transmission time is the same as or before the temporary slot (Node C); and (3) nodes whose schedules are unknown (Node D). In Figure 6.4, the box with solid border indicates the interval that is already occupied by the corresponding node; while the dashed line one means that the node is a potential competitor for these slots. To solve the contention, the standard defines a pseudorandom function with the slot sequence number and all IDs of the competing nodes as inputs. The output values are called mixing values. If the current node ID and the slot number generate the largest mixing value, it wins this slot and broadcasts the new schedule to the neighbors. If the node fails, the next transmission opportunity is set to be the temporary transmission slot, and the above competing procedure is repeated until it wins in a slot. Based on the above description, we can see that the probability of a node winning a contention is determined by the total number of competing nodes, and number of competing nodes is

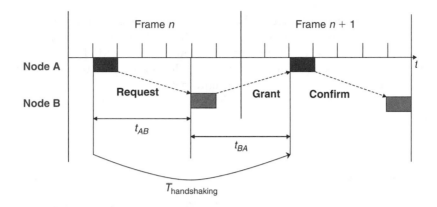

FIGURE 6.5
802.16 mesh three-way handshake.

related with the number of neighbor nodes and their topologies that are also analyzed in Sections 6.3 and 6.4.

The 802.16 distributed mesh scheduler employs a three-way handshaking procedure to set up connections with neighbors. The procedure is shown in Figure 6.5. The requester sends a *request* message in the MSH-DSCH packet along with the data subframe availability information. After receiving the request, the receiver responses with a *grant* message indicating all or a subset of the suggested data subframe. When the requester receives the grant message, it transmits a *confirm* message to the receiver containing a copy of the granted subframe. With this mechanism, the neighbors of both the requester and the receiver can have the up-to-date data subframe allocation information. Since a centralized scheduler is not required to coordinate all the transmissions in the mesh network, distributed scheduling exhibits better flexibility and scalability as a transmission scheduling mechanism in the mesh mode.

Although the messages and signaling mechanisms for transmission scheduling are defined in the IEEE 802.16 mesh mode, how the minislots should be assigned to the different stations is left unspecified. Refs. 7–10 concentrate on the centralized scheduling mechanism for achieving QoS and fairness objectives in the mesh mode. Ref. 11 presented a slot allocation algorithm by prioritization to achieve QoS with distributed scheduling. A combination scheduler of both centralized and distributed manners along with a bandwidth allocation is used to achieve high throughput in Ref. 12. Since the centralized scheduler determines the resource allocation for all the nodes, the system performance of centralized scheduling can be easily analyzed and leveraged. In contrast, in distributed scheduling the channel access control is more complex because an election-based algorithm is used and every node computes its transmission time without global information. It is necessary to understand the distributed scheduler behavior thoroughly to optimize the network throughput and delay performance. In this chapter, we focus on the performance of the distributed-scheduling algorithm. First, we develop a stochastic model to analyze the control channel performance. This model considers the important parameters that could affect the system performance like the total node number, holdoff exponent value, and topology. With this model, the channel contention situation and connection setup time variance can be evaluated clearly under different parameters. We also implement the 802.16 mesh mode in ns-2 simulator, and the theoretical and simulation results match very well.

6.3 Performance Analysis of IEEE 802.16 Distributed Scheduler

In distributed scheduling of IEEE 802.16 mesh mode, the channel contention behavior is correlated with the number of nodes, exponent value, and network topology. In our study, we assume the transmit time sequences of all the nodes in the control subframe form statistically independent renewal processes. Based on this assumption, we develop a stochastic model to estimate the control channel performance.

6.3.1 Model and Approach

The performance metrics of interest in the MAC layer include the throughput and the delay. In the IEEE 802.16 mesh mode, the detail of the data subframe reservation is left unstandardized and is to be implemented by the vendors; and the control subframe is independent of the data subframe. We consider the modeling and analysis of the control subchannel, which is characterized by the distributed election algorithm. Assume that the number of nodes in the network is N. Let $\mathcal{N}_k$ denote the set of 2-hops neighbor nodes of node k, $N_k = |\mathcal{N}_k|$; $\mathcal{N}_k^{\text{unknown}}$ denote the set of nodes whose the schedules are unknown in the neighbor nodes set $\mathcal{N}_k$, $N_k^{\text{unknown}} = |\mathcal{N}_k^{\text{unknown}}|$; $\mathcal{N}_k^{\text{known}} = \mathcal{N}_k \backslash \mathcal{N}_k^{\text{unknown}}$ and $N_k^{\text{known}} = |\mathcal{N}_k^{\text{known}}|$. Let x_k, $k = 1, \dots, N$, denote the holdoff exponent of node k, then $H_k = 2^{x_k+4}$ is the holdoff time of node k; and $V_k = 2^{x_k}$ is the eligibility interval of node k. Let S_k denote the number of slots that node k fails before it wins the first slot with the pseudorandom competition, which is a random variable, then the interval between successive transmission opportunities is $\tau_k = H_k + S_k$. Our goal is to determine the distributions or mean values of τ_k by modeling and analyzing the distributed-scheduling algorithm, based on which the throughput and delay performance metrics can be derived.

Let $Z_k(t)$ denote the number of transmission times of node k up to slot t, then $Z_k(t)$ is a counting process with interevent time τ_k. To simplify the analysis, we make the following assumptions: (1) the counting process of each node eventually reaches its steady state and the intervals are i.i.d., that is, $Z_k(t)$ forms a stationary and ergodic renewal process and (2) the renewal processes of different nodes are mutually independent at their steady states. Note that the distribution of the renewal intervals of each node depends on the number of competing nodes in its neighborhood and their holdoff exponents. However, when all the processes reach their steady states, we can assume that the processes are initiated at $t = -\infty$, and the time of renewal events of different processes are uncorrelated. Our analysis is based on the above assumptions.

Suppose that the expected number of competing nodes in slot s for the node k is $M_k(s)$. As a result of the pseudorandom election algorithm, the probability that this node wins the slot is

$$p_k(s) = \frac{1}{M_k(s)} \tag{6.1}$$

So the probability mass function of S_k is

$$P(S_k = s) = \prod_{i=1}^{s-1} [1 - p_k(i)] p_k(s), \quad s = 1, 2, \ldots \tag{6.2}$$

To get the distribution of τ_k, we need to find $M_k(s)$. But $M_k(s)$ depends on the distributions of τ_j, $j = 1, \ldots, N_k$, since all the nodes in the neighborhood of node k are candidates to compete with node k. In this paper, we take the following approach to solve this problem: we will derive $M_k(s)$ in terms of $\{p_k(s)\}$ by modeling the distributed election algorithm, and then by using the relation $p_k(s) = 1/M_k(s)$, we obtain a set of close form equations for $\{p_k(s)\}$ to solve them.

6.3.2 Collocated Scenario

To simplify the analysis, we first consider the collocated scenario: all nodes are one-hop neighbors of each other. In this simple case, there is no unknown node, and $N_k = N$; $k = 1, 2, \ldots, N$; that is, all nodes have the same neighborhood.

6.3.2.1 Identical Holdoff Exponent

To further simplify the analysis, we first assume equal holdoff exponents, that is $x_1 = x_2 = \cdots = x_N$. Hence when the nodes are collocated, the transmission interval τ_k has the same distribution, $p_k(s) \equiv p(s)$, $s = 1, 2, \ldots, \forall k$. So we can drop the subscript k.

To proceed with the analysis, we need to introduce the notion of excess time of a renewal process. Let $Z(t)$ be a renewal process and t be any chosen time slot, the spread, $\tau_{Z(t)+1}$, is the renewal interval in which t lies, as shown in Figure 6.6. The age of the renewal process, $a(t)$, is the time since the last renewal before t; the excess (or residual life time) of the renewal process, $e(t)$, is the time to the next renewal after t. The limiting distribution of excess time is established by the following lemma, which is a corollary of the *Renewal*

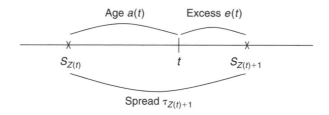

FIGURE 6.6
The age and excess time of a renewal process.

Reward Theorem. The proof is similar to that for the continuous-time version of the lemma in Ref. 13.

LEMMA 6.1
(Limiting distribution of excess time) *Let* $Z(t)$ *be a renewal process and* τ *be the renewal interval and* $e(t)$ *be the excess time, the limiting distribution of the excess time is* $\lim_{t\to\infty} \sum_{s=1}^{t} I\{e(s) \le y\}/t = 1/\mu \sum_{i=1}^{y} P(\tau \ge i)$ *for fixed* $y \ge 0$, *where* $\mu = E[\tau]$ *and* $I\{\cdot\}$ *is an indicator function.*

By the stationary and ergodic assumption, we can take the limiting distribution of excess time as its stationary distribution

$$P(e \le y) = \frac{1}{\mu} \sum_{i=1}^{y} P(\tau \ge i) \qquad (6.3)$$

The distribution of τ is given in terms of $\{p(s)\}$ as

$$P(\tau \ge i) = \begin{cases} 1, & \text{if } i \le H \\ \sum_{s=i-H}^{\infty} \prod_{j=1}^{s-1} [1 - p(j)]p(s), & \text{if } i > H \end{cases} \qquad (6.4)$$

Using Equations 6.3 and 6.4, we have

$$P(e = y) = P(e \le y) - P(e \le y - 1) = \frac{1}{\mu} P(\tau \ge y)$$

$$= \begin{cases} \dfrac{1}{\mu}, & \text{if } y \le H \\ \dfrac{1}{\mu} \sum_{s=y-H}^{\infty} \prod_{j=1}^{s-1} [1 - p(j)]p(s), & \text{if } y > H \end{cases} \qquad (6.5)$$

where $\mu = H + E[S] = H + \sum_{s=1}^{\infty} s \prod_{j=1}^{s-1} [1 - p(j)]p(s)$.

Employing the above results, we can calculate the probability that another node j will compete with node k in the given slot $S_k = s$. In this simplified scenario, there is no unknown node, the possible competing scenario is either

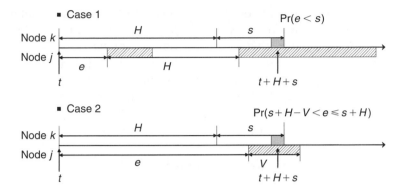

FIGURE 6.7
The competing scenarios of identical holdoff exponent case.

case 1 or case 2, as shown in Figure 6.7. Recall that $H = 2^{x+4}$ is the holdoff time, and $V = 2^x$ the eligibility interval length. Since the renewal processes of node k and node j are assumed to be statistically independent, at the current transmit time t of node k, the time from t to the next transmit time of node j is simply the excess time $e_j(t)$ of node j. By the assumption that the renewal process is stationary and that the distributions of τ_k are identical, we can simply denote $e_j(t)$ as e. The competing probability for case 1 is

$$P(\text{Earliest\_Next\_Xmt\_Time}_j \leq \text{Temp\_Xmt\_Time}_k)$$

$$= P(t + e + H + 1 \leq t + H + s) = P(e < s) \tag{6.6}$$

The competing probability for case 2 is

$$P(\text{Temp\_Xmt\_Time}_k \in \text{Next\_Xmt\_Eligible\_Interval}_j)$$

$$= P(t + e \leq t + H + s \leq t + e + V - 1) = P(s + H - V < e \leq s + H) \tag{6.7}$$

So the probability that node j will compete with node k in the given slot $S_k = s$ is the sum of the probabilities of the above two cases, and by using Equation 6.5 we get

$$P(\text{node } j \text{ competes with node } k | S_k = s)$$

$$= P(e < s | S_k = s) + Pr(s + H - V < e \leq s + H | S_k = s)$$

$$= \begin{cases} \dfrac{V}{\mu} + \dfrac{1}{\mu} \displaystyle\sum_{i=H+1}^{s+H} \sum_{l=i-H}^{\infty} \prod_{j=1}^{l-1} [1 - p(j)]p(l), & \text{if } s \leq V \\[18pt] \dfrac{s}{\mu} + \dfrac{1}{\mu} \displaystyle\sum_{i=s+H-V+1}^{s+H} \sum_{l=i-H}^{\infty} \prod_{j=1}^{l-1} [1 - p(j)]p(l), & \text{if } V < s \leq H \\[18pt] \dfrac{H}{\mu} + \dfrac{1}{\mu} \left(\displaystyle\sum_{i=H+1}^{s} + \sum_{i=s+H-V+1}^{s+H} \right) \cdot \left(\sum_{l=i-H}^{\infty} \prod_{l-1}^{j=1} [1 - p(j)]p(l) \right), & \text{if } s > H \end{cases} \tag{6.8}$$

When the holdoff exponents of all the nodes are identical, this probability is the same for any two nodes k and j. We denote it as $P(\text{compete}|S=s)$, and denote $N^k_{\text{compete}}(s)$ as the number of nodes (among $N-1$ neighbors), which compete with node k in slot s. By the assumption of statistical independence, $N^k_{\text{compete}}(s)$ is binomial distributed, that is, $N^k_{\text{compete}}(s) \sim \mathcal{B}[N-1, P(\text{compete}|S=s)]$. Hence, the expected number of nodes competing with node k in slot s is $E[N^k_{\text{compete}}(s)] = (N-1)P(\text{compete}|S=s)$, and the competing nodes in slot s for node k are

$$M(s) = (N-1)P(\text{compete}|S=s) + 1 \qquad (6.9)$$

Substituting Equation 6.9 into Equation 6.1, we get

$$p(s) = \frac{1}{M(s)} = \frac{1}{(N-1)P(\text{compete}|S=s) + 1} \qquad (6.10)$$

Combining Equations 6.8 and 6.10, we get a set of equations by which we can solve for $p(s)$, $s=1, 2, \ldots$. Typically, $p(s) \to 0$ as $s \to \infty$, so we can truncate the tail and consider only $p(s)$, $s=1, 2, \ldots, L$ for some large L, and then solve the fixed point equations by using standard iterative method. The computation complexity of the above approach is high since there are L elements to update at each iteration, where L should be typically chosen large enough. As we will see in the next subsection, this approach becomes more complicated when the holdoff exponents are not identical. This renders it difficult for performance evaluation and impractical for online performance optimization. We next propose a simplified approach by assuming that S follows a geometric distribution.

By observing the histograms of our simulation data, we find that the distribution of S can be approximated by a geometric distribution. So we make a further approximation that $p(1)=p(2)= \cdots =p$, and

$$P(S=s) = (1-p)^{s-1}p \qquad (6.11)$$

Then we have $E[S] = \frac{1}{p}$. Similar to Equations 6.4 and 6.5, we can derive the distribution of τ as follows:

$$P(\tau \geq y) = \begin{cases} 1, & \text{if } y \leq H \\ (1-p)^{y-H-1}, & \text{if } y > H \end{cases} \qquad (6.12)$$

and the distribution of e as

$$P(e=y) = \frac{1}{\mu}P(\tau \geq y) = \begin{cases} \dfrac{1}{\mu}, & \text{if } y \leq H \\ \dfrac{1}{\mu}(1-p)^{y-H-1}, & \text{if } y > H \end{cases} \qquad (6.13)$$

Then the probability that another node j will compete with node k in the given slot $S_k = s$ is

$$P(\text{node } j \text{ competes with node } k | S_k = s)$$
$$\triangleq P(\text{compete}|S = s)$$
$$= P(e < s|S_k = s) + P(s + H - V < e \le s + H|S_k = s)$$
$$= \begin{cases} \dfrac{V}{\mu} + \dfrac{1 - (1 - p)^s}{\mu p}, & \text{if } s \le V \\[2ex] \dfrac{s}{\mu} + \dfrac{(1 - p)^{s-V} - (1 - p)^s}{\mu p}, & \text{if } V < s \le H \\[2ex] \dfrac{H}{\mu} + \dfrac{1 - (1 - p)^{s-H-1} + (1 - p)^{s-V} - (1 - p)^s}{\mu p}, & \text{if } s > H \end{cases}$$

$$(6.14)$$

For the geometric distribution, $p(s) = p$ for all s, hence $M(s) = M$ for all s, which implies that the number of competing nodes in each slot s should be the same. Here we approximate M as the expectation of $M(s)$ as

$$\frac{1}{p} = M \simeq E_S[M(s)] = E_S[(N - 1)P(\text{compete}|S = s) + 1] \qquad (6.15)$$

Using Equations 6.14 and 6.15 and after some manipulations, we get

$$\frac{1}{p} = (N - 1)\left(\frac{V}{\mu} + \frac{1}{\mu p} - \frac{(1 - p)^{V+1} - (1 - p)^{H+2}}{\mu p(2 - p)}\right) + 1 \qquad (6.16)$$

Note that the last term in the second bracket of RHS is typically small, we can simplify Equation 6.16 by dropping that term as

$$\frac{1}{p} \simeq (N - 1)\left(\frac{V}{\mu} + \frac{1}{\mu p}\right) + 1 \qquad (6.17)$$

By substituting $E[S] = \frac{1}{p}$ and $\mu = H + E[S]$ into Equation 6.17, we express the above equation in terms of $E[S]$ as

$$E[S] = (N - 1)\frac{V + E[S]}{H + E[S]} + 1 = (N - 1)\frac{2^x + E[S]}{2^{x+4} + E[S]} + 1 \qquad (6.18)$$

A fixed point iteration can be used to obtain $E[S]$ from Equation 6.18.

6.3.2.2   Nonidentical Holdoff Exponents

Now we consider the case where holdoff exponents x_k, $k = 1, \ldots, N$ are not identical, but we still restrict ourselves to the collocated scenario. The analysis is similar to the identical holdoff exponent case. The competing scenarios are illustrated in Figure 6.8. The competing probability for case 1 is

$$P(\text{Earliest\_Next\_Xmt\_Time}_j \le \text{Temp\_Xmt\_Time}_k)$$

$$= P(t + e_j + H_j \le t + H_k + s) = P(e_j \le \min\{0, s + H_k - H_j\}) \quad (6.19)$$

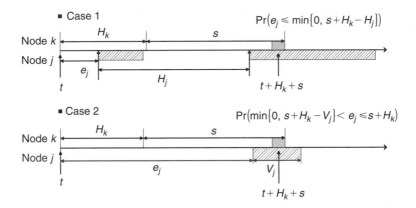

FIGURE 6.8
The competing scenarios of nonidentical holdoff exponents case.

And the competing probability for case 2 is

$$P(\text{Temp\_Xmt\_Time}_k \in \text{Next\_Xmt\_Eligible\_Interval}_j)$$

$$= P(t + e_j \leq t + H_k + s \leq t + e_j + V_j)$$

$$= P(\min\{0, s + H_k - V_j\} < e \leq s + H_k) \tag{6.20}$$

Denote $P(C_k^j | S_k = s)$ as the probability that node j will compete with node k in the given slot $S_k = s$, which is given by

$$P(C_k^j | S_k = s) \triangleq P(\text{node } j \text{ competes with node } k | S_k = s)$$

$$= P(e_j < \min\{0, s + H_k - H_j\}) + P(\min\{0, s + H_k - V_j\}$$

$$< e \leq s + H_k)$$

$$= \begin{cases} P(e_j \leq s + H_k | S_k = s) = \dfrac{s + H_k}{\mu_j}, & \text{if } s + H_k \leq V_j \\[2mm] P(s + H_k - V_j < e \leq s + H_k | S_k = s) = \dfrac{V_j}{\mu_j}, \\[2mm] \quad \text{if } V_j < s + H_k \leq H_j \\[2mm] P(e_j < s + H_k - H_j | S_k = s) + P(s + H_k - V_j < e \leq s \\[1mm] \qquad\qquad\qquad\qquad\qquad\qquad + H_k | S_k = s) = (*), \\[2mm] \quad \text{if } s + H_k > H_j \end{cases}$$

$$\tag{6.21}$$

where $\mu_j = H_j + E[S_j] = H_j + \sum_{s=1}^{\infty} s \prod_{i=1}^{s-1} [1 - p_j(i)] p_j(s)$. The detail expression of $(*)$ is very complex, so we make some simplification here. Again, by the assumption of statistical independence, $M_k(s) = \sum_{j \neq k, j=1}^{N} P(C_k^j | S_k = s) + 1$, and we have

$$p_k(s) = \frac{1}{M_k(s)} = \frac{1}{\sum_{j \neq k, j=1}^{N} P\left(C_k^j | S_k = s\right) + 1} \tag{6.22}$$

Similarly, we can find the distributions of τ_j, $j = 1, 2, \ldots, N$ by the fixed point iterations. We further assume that S_j, $j = 1, 2, \ldots, N$ are geometrical distributed as before, that is, assume $p_j(1) = p_j(2) = \cdots \triangleq p_j$, and

$$P(S_j = s) = (1 - p_j)^{s-1} p_j \tag{6.23}$$

Hence the distribution of e_j is

$$P(e_j = y) = \frac{1}{\mu_j} P(\tau_j \geq y) = \begin{cases} \dfrac{1}{\mu_j}, & \text{if } y \leq H_j \\[2mm] \dfrac{1}{\mu}(1 - p_j)^{y - H_j - 1}, & \text{if } y > H_j \end{cases} \tag{6.24}$$

Then $(*)$ can be simplified as

$$(*) = P(e_j < s + H_k - H_j | S_k = s) + P(s + H_k - V_j < e \leq s + H_k | S_k = s)$$

$$= \begin{cases} \dfrac{V_j}{\mu_j} + \dfrac{1 - (1 - p_j)^{s + H_k - H_j}}{\mu_j p_j}, & \text{if } H_j < s + II_k \leq H_j + V_j \\[4mm] \dfrac{(1 - p_j)^{s + H_k - H_j - V_j} - (1 - p_j)^{s + H_k - H_j}}{\mu_j p_j} + \dfrac{s + H_k - H_j}{\mu_j}, \\[2mm] \quad \text{if } H_j + V_j < s + H_k \leq 2H_j \\[4mm] \dfrac{H_j}{\mu_j} + \dfrac{1 - (1 - p_j)^{s + H_k - 2H_j} + (1 - p_j)^{s + H_k - H_j - V_j}}{\mu_j p_j} \dfrac{(1 - p_j)^{s + H_k - H_j}}{\mu_j p_j}, \\[2mm] \quad \text{if } s + H_k > 2H_j \end{cases} \tag{6.25}$$

To estimate p_k, we proceed as follows:

$$\frac{1}{p_k} = M \simeq E_{S_k}[M_k(s)] = \sum_{j \neq k, j=1}^{N} E_{S_k}\left[P\left(C_k^j | S_k = s\right)\right] + 1 \tag{6.26}$$

We can derive $E_{S_k}[P(C_k^j | S_k = s)]$ from Equations 6.21, 6.23, and 6.25. The formulation is very complex and takes different forms depending on the value of x_k and x_j. Similarly, we can make some approximations to find the counterparts of Equation 6.18 as in the last subsection. Here we directly give the approximation result as

$$E_{S_k}\left[P\left(C_k^j | S_k = s\right)\right] \simeq \begin{cases} \dfrac{V_j + E[S_k]}{\mu_j}, & \text{if } x_j \geq x_k \\[2mm] 1, & \text{if } x_j < x_k \end{cases} \tag{6.27}$$

Note that $E[S_k] = \frac{1}{p_k}$, combining Equations 6.26 and 6.27, we have

$$E[S_k] = \sum_{j=1, j \neq k, x_j \geq x_k}^{N} \frac{2^{x_j} + E[S_k]}{2^{x_j+4} + E[S_j]} + \left(\sum_{j=1, j \neq k, x_j < x_k}^{N} 1 \right) + 1, \quad k = 1, \ldots, N$$

(6.28)

Again we can solve for $E[S_k]$ using fixed point iteration.

6.3.3 General Topology Scenario

Now we extend the results to the general topology scenario. The difference between the collocated scenario and the general topology scenario is that in the general topology: (1) the neighborhood node sets of different nodes $\mathcal{N}_k$ may be different; (2) there exist unknown nodes, hence besides the competing cases 1 and 2, there are unknown competing nodes. However, the analysis procedure is almost the same as above, Equation 6.26 still holds with minor modifications

$$E[S_k] = \frac{1}{p^k} \simeq \sum_{j \neq k, j=1}^{N_k} E_{S_k}\left[P\left(C_k^j | S_k = s \right) \right] + 1$$

(6.29)

where $N_k = N_k^{\text{known}} + N_k^{\text{unknown}}$. Since unknown nodes are always regarded as competing nodes in each slot by the distributed election algorithm, we have

$$Pr\left(C_j^k | S_k = s \right) = P(j \text{ compete with } k \text{ in slot } s | S_k = s)$$

$$= 1, \quad j \in \mathcal{N}_k^{\text{unknown}}, \quad \forall s$$

(6.30)

Taking unknown nodes into account, we find the expression for $E[S_k]$ by a similar procedure as in the last section.

$$E[S_k] = \sum_{j=1, j \neq k, x_j \geq x_k}^{N_k^{\text{known}}} \frac{2^{x_j} + E[S_k]}{2^{x_j+4} + E[S_j]} + \left(\sum_{j=1, j \neq k, x_j < x_k}^{N_k^{\text{known}}} 1 \right) + N_k^{\text{unknown}} + 1,$$

$$k = 1, \ldots, N$$

(6.31)

Again we can solve for $E[S_k]$ by fixed point iterations.

6.3.4 Performance Metrics Estimation

Let $T_{\text{handshake}}^{AB}$ denote the time node; A needs to accomplish a three-way handshaking with node B. Now we can derive $E[T_{\text{handshake}}^{AB}]$, given the distributions of τ_k of all nodes $k = 1, \ldots, N$. From the procedure of the three-way handshaking as illustrated in Figure 6.5, we can see that $T^{AB}_{\text{handshake}} = t_{AB} + t_{BA}$, where t_{AB} is the time interval between node A sending a request to B and B replying a grant to A, and t_{BA} is the time interval between B sending the grant and A replying with a confirm. By the independence assumption, and further assuming that the renewal processes of node A and B have run for a long time, we can assume that t_{AB} follows a limiting distribution as the excess time e_B. However, t_{BA} may not follow the same distribution as e_A since it is dependent on t_{AB}, thus the limiting distribution does not hold. It is very complicated to find the exact distribution of t_{BA} since the renewal interval τ is not geometrically distributed. Hence we take an empirical approach here. Note that when $\tau_A \ll \tau_B$, the renewal process of node A can be considered to run for a long time after sending *request*, so in this case we can assume that $t_{BA} \simeq e_A$. In this case, $E[T^{AB}_{\text{handshake}}] \simeq E[e_B] + E[e_A]$. However, when $\tau_A \gg \tau_B$, $t_{BA} \simeq \tau_A$ since the portion of excess time e_B is negligible, and $E[T^{AB}_{\text{handshake}}] \simeq E[e_B] + E[\tau_A] = E[e_B] + \mu_A$. Empirically, we can estimate $E[T^{AB}_{\text{handshake}}]$ as

$$E\left[T^{AB}_{\text{handshake}}\right] \simeq E[e_B] + \alpha \mu_A + (1 - \alpha) E[e_A] \qquad (6.32)$$

where α is a compromising factor and $\mu_A = E[\tau_A] = H_A + E[S_A]$. Based on the simulation data, we find $\alpha = E[S_A]/\mu_A$ to be a good choice for the identical holdoff exponent case and $\alpha = [\mu_A/(\mu_A + \mu_B)]^2$ for the nonidentical holdoff exponents case. Once $E[S_k]$ is known, from Equation 6.25 we can calculate $F[e_k]$ as

$$E[e_k] = \sum_{y=1}^{H_k} \frac{y}{\mu_k} + \sum_{y=H_k+1}^{\infty} \frac{y}{\mu_k}(1 - p^k)^{y-H-1}$$

$$= \frac{H_k(H_k + 1)}{2\mu_k} + \frac{H_k}{\mu_k p^k} + \frac{1}{\mu_k p^{k^2}}$$

$$= \frac{2^{x_k+3} + 2^{2x_k+7} + 2^{x_k+4}E[S_k] + E^2[S_k]}{2^{x_k+4} + E[S_k]} \qquad (6.33)$$

Employing the results of $E[S_k]$ from previous subsections, we can determine $E[T^{AB}_{\text{handshake}}]$ for any two neighbor nodes A and B by using Equations 6.32 and 6.33.

6.4 Evaluation

In this section, we provide the ns-2 simulation results for various scenarios and compare them with the analytical ones obtained by the modeling framework developed in this paper.

6.4.1 Simulation Methodology

The typical ns-2 simulator includes typical TCP/IP/LL/MAC/PHY stack. The current MAC modules for ns-2 include 802.11, ethernet, TDMA and satellite; however, no 802.16 MAC module is available. In our work, we implement a new MAC module for the IEEE 802.16 mesh mode and use it to evaluate the system performance. The module consists of three logic parts, the network controller, the scheduling controller, and the data channel management component. The network controller is responsible for the network configuration, node entry, synchronization, etc. The scheduling controller part handles the control channel contention, three-way handshaking, and the data channel allocation. During the holdoff time of a node, the MSH-DSCH messages received from PHY module are sent to the scheduling controller so that the scheduling information in two-hop neighborhood is updated. In the transmission slot, the scheduling controller contends the next transmission time using the election algorithm based on the collected information. Once the node wins the contention, a MSH-DSCH message is sent to the PHY layer. And a new timer is set to activate the next transfer in the next transmission time. Besides the scheduling, the scheduling controller also sets the request field properly in the MSH-DSCH message if data packets are received from LL layer and replies with grant or confirm according to the three-way handshaking procedure. The data channel component receives and transmits data packets in the allocated time slots. In our simulation, we use a simple reservation mechanism by assigning each data packet in one minislot.

The exponent value determines a node eligible-interval length and the channel contention. In our simulations, the set of possible exponent values is {0, 1, 2, 3, 4}. During initialization, every node is assigned an ID randomly, and the initial transmission time for each node is arranged sequentially. In the mesh mode, the node IDs are the input parameters of the election algorithm. The randomness of the node IDs makes sure that the initial transmission time arrangement does not affect the final results. In our simulation, we investigate two types of network topologies, namely, collocated and general topologies, respectively.

1. *Collocated scenario*: In the collocated scenario, all nodes are within the radio transmission range of each other so that every node can obtain the up-to-date schedule information of all the neighbors. The IEEE 802.16 is a new standard and lacks field data. The total transmission opportunities in control channel becomes 256 when the exponent value is 4. In choosing the exponent value, we believe 4 is large enough; otherwise, the connection setup latency will become too long to tolerate. In our simulation, both the identical and non-identical cases are studied. Besides the exponent values, we also vary the total node number from 10 to 128 to study the scheduler performance.

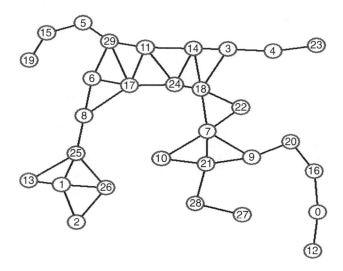

FIGURE 6.9
The node placement in general topology.

2. *General topology*: In the general topology, the scheduling information of some neighbors beyond one hop may become stale. The reason is that some one-hop neighbors could not update the schedule information in time for some specific transmission order. For example, in Figure 6.9, if node 19 transmits first and node 5 transmits second, node 5 does not know the new schedule information of node 19 during the contention because node 15 has not updated it yet. During the transmission time, a node can tell if a neighbor's schedule information is stale or not by comparing the node's schedules with the current time. A neighbor node becomes unknown node once its schedule information is stale and will always be treated as potential competitor (node D in Figure 6.4). So, in the general topology, the potential competing node's number for a node differs from that in the collocated scenario.

In our simulation, the node placement of general topology is shown in Figure 6.9. There are totally 30 nodes and the one-hop neighbors are connected by lines. The exponent values for the nodes are not identical, and are also assigned arbitrarily from set {0, 1, 2, 3, 4}. In this topology, we assign larger exponent values to the nodes residing between two nodes with smaller exponent values. The node with larger exponent value has longer update period, so that the scheduling information of the two separated nodes are apt to become stale. In simulation, if the current node finds that the stored schedule information of a neighbor is out-of-date, it treats the neighbor as an unknown node and counts it as a competitor.

6.4.2 Numerical Results

6.4.2.1 Transmission Interval

In the identical exponent case, all nodes have the same expected holdoff time. However, the actual holdoff time depends on the channel contention situation. Figure 6.10 shows the expected node consecutive transmission intervals $H + E[S]$ obtained from Equation 6.18 and simulations under various node numbers and exponent values, where N is the total node number and x the exponent value. We can see that the two sets of results match well. In the figure, the node transmission intervals increase with the total node number and exponent values. When N becomes larger, the contention becomes more intensive so that a node has to compete more times before it wins. With the increase of the exponent value, node holds off longer time before the next transmission, therefore the intervals increase with x too. In the nonidentical exponent case, we divide the N nodes into five groups of equal size. The nodes in each group take the same exponent value from the set $\{0, 1, 2, 3, 4\}$. The analytical and simulation results are also compared. The figure is not shown because of the space limit and the detail can be found in Ref. 14. In this case, the system behaves similarly as in the identical exponent scenario with longer transmission intervals. The excellent match between the two sets of results demonstrates that our model is very accurate.

In the 802.16 mesh mode, the interval between two consecutive transmissions is critical to the system performance. From the above results, we can conclude that the interval increases with the total neighbor number and exponent value. If there are more nodes in the neighborhood, the contention becomes intensive, so a node needs more time to get access to the channel. In the figures, we can also see that, for different values of N, when $x = 0$, the competition is the most severe and a node fails many times before winning,

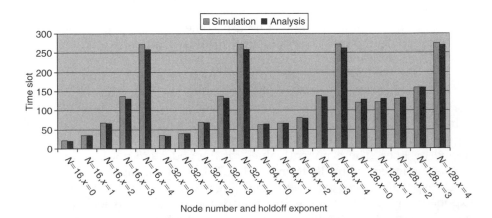

FIGURE 6.10
Simulation and analytical results on the expected transmission intervals for the identical exponent.

especially when N is large. With the increase of the exponent value, the contention becomes less competitive because nodes have longer holdoff time; however, the transmission interval becomes longer too. We can also conclude that the exponent values have more significant impact on the system performance. Based on the results, a node with real-time traffic should have smaller exponent value so that it can have more chance to access the channel. In contrast, too many nodes with small exponent values will generate serious contention. Hence, an 802.16 system can be optimized by assigning appropriate exponent values to the nodes in the network.

6.4.2.2 Three-Way Handshaking Time

The IEEE 802.16 mesh mode employs a three-way handshaking procedure to set up connection. In this section, we compare the theoretical connection setup time with that got by simulations. In our simulation, the pairs of requester and grantor are selected arbitrarily and the results are the average of the data obtained from all pairs.

The three-way handshaking time for the identical exponent scenario is shown in Figure 6.11. In the simulation, the total node numbers are set to be 10, 30, 50, and 100, respectively. The solid lines represent the theoretical

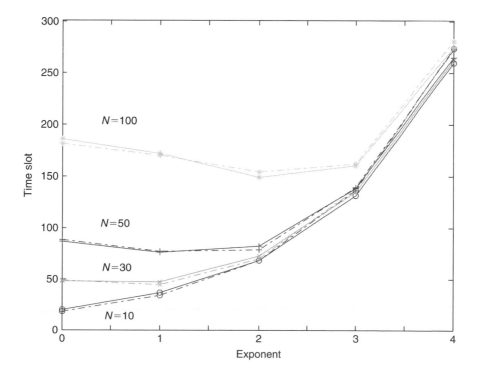

FIGURE 6.11
Simulation and analytical results on the three-way handshaking time for the identical exponent case.

results and the dotted lines are the data obtained by simulations. We can see that the two sets of results match very well, and the connection setup time increases with the total node number and the exponent values. When N is large enough ($N = 100$), the three-way handshaking time decreases first with the increase of exponent value and then goes up again. The reason is that the increase of x can reduce the channel competition to some extent when N is large. This phenomenon verifies our conclusion on the transmission interval. However, when the exponent x becomes large enough ($x = 4$), the holdoff time dominates, hence, the connection setup time becomes similar under different number of nodes.

The three-way handshaking time for the nonidentical exponent scenario is also studied. Again we set the total node number to be 10, 30, 50, and 100, respectively. The total nodes are divided into five groups of equal size. The nodes in each group take the same exponent value from the set {0, 1, 2, 3, 4}. Here, we present the case for $N = 100$ (Figure 6.12) only. In these figures, the five pairs of curves are the connection setup time whose requesters have exponents 0, 1, 2, 3, and 4, respectively. The abscissa represents the grantor exponent values. The solid lines are the analytical results and the dotted lines are the simulation data. Again, our model is very accurate for the nonidentical exponent scenario as well.

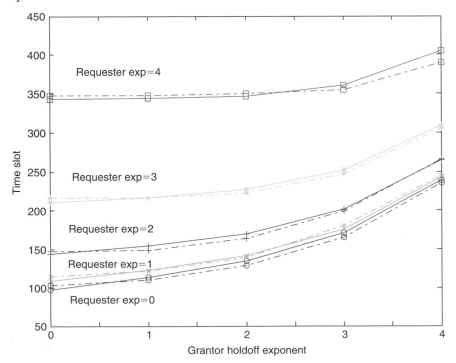

FIGURE 6.12
Simulation and analytical results on the three-way handshaking time for the nonidentical exponent case with $N = 100$.

Different from the identical exponent case, the connection setup time for the nonidentical case is related to the requester and grantor order besides the exponent values. If the node with smaller exponent value is the requester, the requester node has more chances to send confirmation message during the grantor's holdoff time after its reply. So the connection setup time is shorter. If the requester exponent value is larger, the requester needs to hold off longer time before sending the confirmation message. In this case, the connection setup time is mainly decided by the requester's holdoff time, and the connection setup time becomes longer too. In the figures, we can also see that the connection setup time increases with the exponent values of the requester and grantor. The connection setup time also increases with the total node number. The curves of requesters with large exponent values are flatter, the reason is that the connection setup time is mainly determined by their eligible interval and holdoff time.

6.4.2.3 General Topology Scenario

The exponent values of some nodes in Figure 6.9 are shown in Table 6.1. The node transmission intervals obtained by simulation and Equation 6.31 are also included. The connection setup times of these nodes are listed in Table 6.2. We can see the simulation and theoretical results are close. With careful study, we can find that the transmission intervals and the connection setup time obtained in general topology are similar to those in collocated scenario. The reason is that although unknown nodes are always treated as potential competitors in general topology, they can also be competing nodes

TABLE 6.1

Simulation and Analytical Results on the Expected Transmission Intervals for the General Topology

Node	Exponent	Simulation	Analysis	Error (%)
0	0	17.97	18.14	0.93
2	2	69.83	67.19	3.78
5	1	36.7	36.8	0.32
6	2	71	69.26	2.45
7	4	275.25	265.35	3.6
8	4	274.7	265.25	3.44
12	1	34.75	34.08	1.93
14	4	273.89	264.45	3.45
16	4	272.27	260.08	4.48
17	3	138.77	134.47	3.1
20	1	35.8	35.68	0.33
21	0	20.08	21.05	4.84
24	1	36.74	37.24	1.36
25	0	18.37	19.38	5.48
26	1	34.82	34.47	1.00
29	4	273.3	263.24	3.68

TABLE 6.2

Simulation and Analytical Results on the Expected Connection Setup Time for the General Topology

Source	Destination	Simulation	Analysis	Error (%)
5	29	156.91	151.63	3.36
14	24	273.83	253.38	7.47
6	17	106.38	107.08	0.66
25	8	149.73	143.7	4.03
2	26	69.83	66.37	4.95
7	21	275.38	258.16	6.25
20	16	154.75	149.29	3.53
0	12	26.83	28.23	5.22

in the collocated topology. The small variance of the total competing node number will not affect the system performance significantly.

6.5 Conclusion

In this chapter, we have developed an analytical model for the distributed scheduling mechanism in the IEEE 802.16 mesh mode. In the mesh mode, every node competes for the channel access in a distributed manner and tries to broadcast its scheduling information periodically. The channel contention result is correlated with the total node number, exponent value, and network topology. Our model assumes that the transmit time sequences of all the nodes in the control subframe form statistically independent renewal processes. Based on this assumption, we have developed methods for estimating the distributions of the node transmission interval and connection setup delay, which are instrumental for evaluating upper layer performance like throughput and delay. To verify the model, we have also implemented the 802.16 mesh MAC module in simulator ns-2. The comparison of the theoretical and simulation results shows that our model is quite accurate under various scenarios.

Since the detail reservation scheme of the data subframe in the IEEE 802.16 mesh mode is left unstandardized, our model sheds some light on the reservation scheme design. For example, based on our analysis, the nodes with real-time traffic shall have smaller holdoff exponents because they can have more chances to obtain data channel. On the other side, too many nodes with small exponent value generate intensive competition that wastes system resource. Then nodes can adjust their exponent values adaptively according to the current competing node number and the traffic load to meet the connection QoS requirements. A good reservation scheme should guarantee the bandwidth allocation fairness and improve the channel utilization at the same time.

Such a reservation scheme needs the information such as the connection setup time and contention success probability provided by our model.

References

1. H. Sari. Trends and challenges in broadband wireless access. *Symposium on Communications and Vehicular Technology (SCVT)*, Oct. 2000, pp. 210–214.
2. IEEE Std 802.16-2004. *IEEE Standard for Local and Metropolitan Area Networks Part 16: Air Interface for Fixed Broadband Wireless Access Systems.* Oct. 1, 2004.
3. C. Eklund, R. B. Marks, K. L. Stanwood, and S. Wang. IEEE standard 802.16: a technical overview of the WirelessMAN™ air interface for broadband wireless access. *IEEE Communication Magazine*, vol. 40, pp. 98–107, June, 2002.
4. Nokia White Paper. *Nokia Rooftop Wireless Routing*. June, 2003.
5. IEEE P802.16a/D3-2001: *Draft Amendment to IEEE Standard for Local and Metropolitan Area Networks—Part 16: Air Interface for Fixed Broadband Wireless Access Systems—Medium Access Control Modifications and Additional Physical Layers Specifications for 2–11 GHz*. Mar. 25, 2002.
6. http://wirelessman.org/tga/contrib/C802.16a-02_30r1.pdf.
7. H. Shetiya and V. Sharma. Algorithms for routing and centralized scheduling to provide QoS in IEEE 802.16 mesh networks. *1st ACM Workshop on Wireless Multimedia Networking and Performance Modelling (WMuNeP'05)*, Montreal, Oct. 2005.
8. H-Y. Wei, S. Ganguly, R. Izmailov, and Z. J. Haas. Interference-aware IEEE 802.16 WiMax mesh networks. *61st IEEE Vehicular Technology Conference (VTC 2005 Spring)*, Stockholm, Sweden, May 29–June 1, 2005.
9. D. Kim and A. Ganz. Fair and efficient multihop scheduling algorithm for IEEE 802.16 BWA systems. *IEEE Broadnets'05*, Boston, MA, 2005.
10. M. Cao, V. Raghunathan, and P. R. Kumar. A tractable algorithm for fair and efficient uplink scheduling of multi-hop WiMax mesh networks. *Proceedings of WiMesh 2006: Second IEEE Workshop on Wireless Mesh Networks*, Reston, VA, 2006.
11. F. Liu, Z. Zeng, J. Tao, Q. Li, and Z. Lin. *Achieving QoS for IEEE 802.16 in Mesh Mode*. School of Electronics and Information Engineering, Tongji University, China, 2005.
12. J. Cheng, C. Chi, and Q. Guo. A bandwidth allocation model with high concurrence rate in IEEE 802.16 mesh mode. *Communications of 2005 Asia-Pacific Conference*, Oct. 2005, pp. 750–754.
13. S. Karlin and H. M. Taylor. *A First Course in Stochastic Processes*. Academic Press, NY, 1975.
14. M. Cao, W. Ma, Q. Zhang, X. Wang, and W. Zhu. Modelling and performance analysis of the distributed scheduler in IEEE 802.16 mesh mode. *ACM MobiHoc 2005*, pp. 78–89.
15. Intel White Paper. *IEEE 802.16 and WiMAX: Broadband Wireless Access for Everyone*. July, 2003.
16. S.-M. Cheng, P. Lin, D.-W. Huang, and S.-R. Yang. A study on distributed/centralized scheduling for wireless mesh network. *ACM International Wireless*

Communications and Mobile Computing Conference (IWCMC'06), Vancouver, WA, July 2006.

17. G. Chu, D. Wang, and S. Mei. A QoS architecture for the MAC protocol of IEEE 802.16 BWA system. *IEEE International Conference on Communications Circuits & System and West Sino Expositions*, vol. 1, China, 2002, pp. 435–439.

18. K. Wongthavarawat and A. Ganz. IEEE 802.16 based last mile broadband wireless military networks with quality of service support. *IEEE Milcom 2003*, vol. 2, 2003, pp. 779–784.

7

Performance Analysis and Simulation Results under Mobile Environments

Mishal Algharabally and Pankaj Das

CONTENTS

7.1 Introduction

Orthogonal frequency division multiplexing (OFDM) is recently becoming very popular, and has already been chosen as the transmission method for many wireless communications standards, such as the IEEE 802.11 wireless local area networks (WLAN) and IEEE 802.16 wireless metropolitan area networks (WMAN) also known as WiMAX. The basic principle of OFDM is to divide the available broadband wireless channel into a large number of narrowband orthogonal subcarriers. This enables OFDM systems to combat the effects of the channel frequency selectivity with low receiver complexity. Unfortunately, OFDM is very sensitive to time variations of the wireless channel. Channel time variations destroy the orthogonality of the OFDM subcarriers and cause energy to leak from one subcarrier to the adjacent subcarriers. The intercarrier interference (ICI) caused by this energy leakage severely degrades the performance of OFDM, and introduces an irreducible error floor.

The WMAN-OFDM PHY (IEEE 802.16-2004) is based on OFDM modulation and designed for non-line-of-sight (NLOS) operation in the frequency bands below 11 GHz [1], and it is very important to understand the abilities and limitations of OFDM systems by investigating their performance under different transmission conditions.

This chapter begins with a detailed description of the system model. The various parts of the uncoded OFDM system are explained in Section 7.2. In Section 7.3, the performance analysis of the OFDM system discussed in Section 7.2 is analyzed. In Section 7.3.1, the performance of the system is analyzed assuming a single antenna at both the transmitter and the receiver. It will be shown that the time variation of the wireless channel will destroy the orthogonality between OFDM subcarriers. This loss of orthogonality will cause energy to leak from one subcarrier to the adjacent subcarriers, and hence, ICI will be present at the receiver. A general case where the receiver does not have a perfect estimate of the wireless channel is considered. The effect of both ICI and channel estimation errors on the quality of the transmission is quantified by deriving a compact expression for the bit-error probability (BEP). In Section 7.3.2, the system model is extended to include space time coding as described in Section 8.3.8 of the IEEE 802.16-2004 standard documentation [1]. It will be shown that in addition to ICI, inter-antenna interference (IAI) caused by the loss of orthogonality of the space time code will also be present, hence, the conventional detection scheme will fail to detect the transmitted OFDM signal, and an alternative detection method that will eliminate IAI and significantly reduce ICI is proposed. Finally, Section 7.4 contains various performance plots obtained by both numerical and simulation analysis of the systems discussed in Sections 7.3.1 and 7.3.2.

7.2   System Model

Figure 7.1 shows the block diagram of a conventional uncoded OFDM system. The input bit stream is first grouped into blocks of size $\log_2(M)$, and then sent to the subcarrier modulator to produce $X(k)$. $X(k)$ is the complex valued data symbol in frequency domain that is modulated onto the kth subcarrier. $X(k)$ may take any value from an M-dimensional constellation depending on the type of mapping used, such as BPSK, QPSK, or QAM for coherent demodulation, or DPSK for noncoherent demodulation. In the IEEE 802.16-2004 standard, the data modulation of choice is M-QAM, where M stands for the number of symbols in the constellation map. Typical values for M are 16, 64, and 256, depending on the required data rate and channel conditions. A block of N complex valued data symbols $\{X(k)\}_{k=0}^{N-1}$ are grouped and converted into parallel (column vector) to form the input to the OFDM modulator. The OFDM modulator consists of an inverse

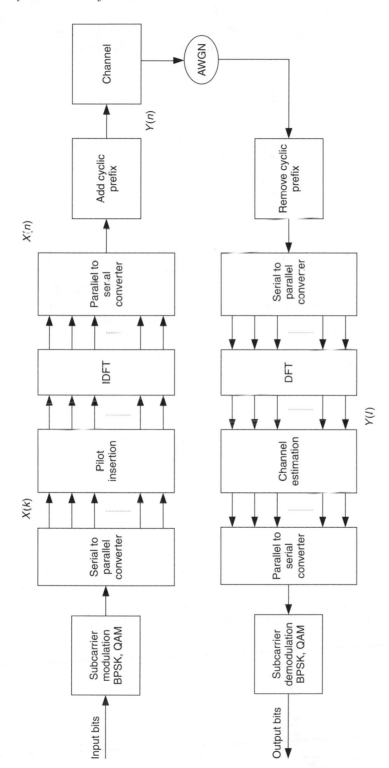

FIGURE 7.1
Uncoded OFDM system.

discrete Fourier transform (IDFT) block, the output of the IDFT block is the modulated OFDM symbol in time domain and is represented by

$$x(t) = \frac{1}{N} \sum_{k=0}^{N-1} X(k) e^{j2\pi f_k t}$$

$$x(n) = x(t)|_{t=\frac{nT}{N}} = \frac{1}{N} \sum_{k=0}^{N-1} X(k) e^{\frac{j2\pi f_k nT}{N}}$$

$$= \frac{1}{N} \sum_{k=0}^{N-1} X(k) e^{\frac{j2\pi nk}{N}} \quad 0 \le n \le N-1 \tag{7.1}$$

where f_k is the frequency of the kth subcarrier, T is the symbol duration, and $j = \sqrt{-1}$. After the OFDM modulator, the modulated OFDM symbol $x(n)$ is parallel to serial converted, and the last G samples of $x(n)$ are copied to the beginning to form the guard interval, which protects the received OFDM symbol from intersymbol interference (ISI) caused by the channel frequency selectivity (delay spread). To ensure that the guard interval will eliminate ISI, the duration of G is chosen to be no less than the channel delay spread. The modulated OFDM symbol after adding the guard interval is

$$x^g(n) = \frac{1}{N} \sum_{k=0}^{N-1} X(k) e^{\frac{j2\pi nk}{N}} \quad G \le n \le N-1 \tag{7.2}$$

$x^g(n)$ is then upconverted before being sent through the frequency selective and time-variant channel. As a practical example from the IEEE 802.16-2004, the following parameters are used:

- Number of subcarriers $N = 256, 1024$
- Guard length $G = 1/4, 1/8, 1/16$, and $1/32$ of the total number of subcarriers

The Stanford University Interim (SUI) channel model consists of a set of six empirical channels that cover most of the terrain types that are typical of the continental United States. The definition of each of the six channels is given in Ref. 2. Table 7.1 shows the parameters for SUI-6 channel.

TABLE 7.1

SUI-6 Channel Model

	Tap 1	Tap 2	Tap 3	Units
Delay	0	14	20	μs
Power	0	−10	−14	dB
Doppler	0.4	0.3	0.5	Hz

Each channel can be modeled as a tapped delay line with time-variant tap gains $h(l, n)$, where l and n are the path and time indices, respectively. The tap gains $\{h(l, n)\}_{l=0}^{L-1}$ are complex Gaussian random variables ($h(l, n) = |h(l, n)|e^{j\phi(n)}$) that are independent and uncorrelated for different values of l; this can be achieved by assuming the wide-sense stationary uncorrelated scattering channels (WSSUS).

If the length of the guard interval G is chosen such that the ISI is negligible, then the received baseband OFDM symbol after removing the first G samples is given by

$$y(n) = \sum_{l=0}^{L-1} h(l, n) x(n - l) + w(n) \tag{7.3}$$

where $w(n)$ is a complex AWGN with one-sided power spectral density N_0.

7.3 Performance Analysis

7.3.1 Single-Input-Single-Output (SISO) Systems

At the receiver, the received symbol is first demodulated by the N-point discrete Fourier transform (DFT), and the demodulated OFDM symbol at the ith subcarrier is given by

$$Y(i) = \sum_{n=0}^{N-1} y(n) e^{\frac{-j2\pi ni}{N}}$$

$$= \sum_{n=0}^{N-1} \sum_{l=0}^{L-1} h(l, n) x(n - l) e^{\frac{-j2\pi ni}{N}} + \sum_{n=0}^{N-1} w(n) e^{\frac{-j2\pi ni}{N}}$$

$$= \sum_{n=0}^{N-1} \sum_{l=0}^{L-1} h(l, n) \frac{1}{N} \sum_{k=0}^{N-1} X(k) e^{\frac{j2\pi(n-l)k}{N}} e^{\frac{-j2\pi ni}{N}} + W(i)$$

$$= \frac{1}{N} \sum_{k=0}^{N-1} X(k) \sum_{n=0}^{N-1} \sum_{l=0}^{L-1} h(l, n) e^{\frac{j2\pi(k-i)n}{N}} e^{\frac{-j2\pi lk}{N}} + W(i) \tag{7.4}$$

Splitting Equation 7.4 into two parts, namely, for $i = k$ and $i \neq k$, respectively, the received OFDM symbol at the kth subcarrier can be represented in the following compact form:

$$Y(k) = \frac{1}{N} \sum_{n=0}^{N-1} H(k,n)X(k)$$

$$+ \frac{1}{N} \sum_{n=0}^{N-1} \sum_{i=0,i\neq k}^{N-1} H(i,n)X(i)e^{\frac{j2\pi(i-k)n}{N}} + W(K)$$

$$= \alpha(k)X(k) + \beta(k) + W(k) \tag{7.5}$$

where $H(k,n)$ is the time-varying frequency response of the channel. The first term above represents the desired signal, which consists of the transmitted symbol $X(k)$ multiplied by the multiplicative distortion $\alpha(k)$. $\beta(k)$ is the ICI owing to the loss of subcarrier orthogonality and $W(k)$ is the Fourier transform of $w(n)$. To show how the expression of the received signal will differ if the channel is time invariant, that is, how does the channel time variation cause ICI, let us start from the orthogonality principle,

$$\frac{1}{N} \sum_{n=0}^{N-1} e^{\frac{j2\pi(u-v)n}{N}} = \begin{cases} 1 & u = v \\ 0 & u \neq v \end{cases} \tag{7.6}$$

Now recall the expression of the ICI

$$\beta(k) = \frac{1}{N} \sum_{n=0}^{N-1} \sum_{i=0,i\neq k}^{N-1} H(i,n)X(i)e^{\frac{j2\pi(i-k)n}{N}}$$

if the channel is time invariant, the above expression reduces to

$$\beta(k) = \sum_{i=0,i\neq k}^{N-1} H(i)X(i)\frac{1}{N} \sum_{n=0}^{N-1} e^{\frac{j2\pi(i-k)n}{N}}$$

and from Equation 7.6, it can be clearly seen that the ICI term vanishes, and the channel will only induce a multiplicative distortion. To gain more insight into the system, it is sometimes desirable to express the system in matrix form. The received signal (Equation 7.3) can be expressed in matrix form as such

$$\mathbf{y} = \mathbf{hx} + \mathbf{w} \tag{7.7}$$

where boldface letters represent vector or matrix quantities. The variables in Equation 7.7 are defined as follows:

$$\mathbf{y} = [y(0)\ y(1) \cdots y(N-1)]^{\mathrm{T}}$$

$$\mathbf{x} = [x(0)\ x(1)\ \cdots\ x(N-1)]^{\mathrm{T}}$$

$$\mathbf{w} = [w(0)\ w(1)\ \cdots\ w(N-1)]^{\mathrm{T}}$$

and the channel matrix **h** is given by

$$
\mathbf{h} = \begin{bmatrix}
h(0,0) & 0 & \cdots & 0 & h(0,L-1) & \cdots & & h(0,1) \\
h(1,1) & h(1,0) & 0 & \cdots & 0 & \ddots & & \vdots \\
\vdots & \ddots & \ddots & \ddots & \ddots & & \ddots & h(L-2,L-1) \\
h(L-1,L-1) & \ddots & \ddots & h(L-1,0) & 0 & & \ddots & 0 \\
0 & h(L,L-1) & \ddots & & \ddots & h(L,0) & & \ddots & \vdots \\
\vdots & \ddots & \ddots & \cdots & & \ddots & & \ddots & 0 \\
0 & \cdots & 0 & h(N-1,L-1) & & \cdots & h(N-1,1) & h(N-1,0)
\end{bmatrix}
$$

$$(7.8)$$

A very important observation here is that the channel matrix **h** is not circulant, which is a direct consequence of the channel time variation. A desirable property of circulant matrices is that they can be diagonalized by pre- and postmultiplication with a unitary matrix. Let F be the DFT matrix defined as

$$\mathbf{F} = [F_{p,q}]_{N \times N} \quad F_{p,q} = \frac{1}{\sqrt{N}} e^{\frac{-j2\pi pq}{N}} \tag{7.9}$$

Applying the DFT to the received OFDM vector in Equation 7.7 we get

$$\mathbf{Y} = \mathbf{F}\mathbf{y} = \mathbf{F}(\mathbf{h}\mathbf{x} + \mathbf{w})$$

$$= \mathbf{F}\mathbf{h}\mathbf{x} + \mathbf{F}\mathbf{w} \tag{7.10}$$

From Equation 7.1, we have $\mathbf{x} = \mathbf{F}^H\mathbf{X}$, where $\mathbf{F}^H$ is the IDFT matrix. The received OFDM vector then becomes

$$= \mathbf{F}\mathbf{h}\mathbf{F}^H\mathbf{X} + \mathbf{F}\mathbf{w}$$

$$= \mathbf{H}\mathbf{X} + \mathbf{W} \tag{7.11}$$

where $(.)^H$ is the conjugate transpose (Hermitian) operation and $\mathbf{H} = \mathbf{F}\mathbf{h}\mathbf{F}^H$ is given by

$$\mathbf{H} = [H_{p,q}]_{N \times N} \quad H_{p,q} = \frac{1}{N} \sum_{n=0}^{N-1} H(q,n) e^{\frac{j2\pi(q-p)n}{N}} \tag{7.12}$$

If the channel is time invariant, then **h** has a circulant structure, and since the DFT matrix **F** is a unitary matrix, this implies that $\mathbf{H} = \mathbf{FhF}^H$ is a diagonal matrix. It can now be stated that mathematically, the ICI is caused by the nondiagonal elements of **H**. The received OFDM vector is now given as

$$
\begin{bmatrix} Y(0) \\ \vdots \\ Y(N-1) \end{bmatrix} = \begin{bmatrix} H_{0,0} & \cdots & H_{0,N-1} \\ \vdots & \ddots & \vdots \\ H_{N-1,0} & \cdots & H_{N-1,N-1} \end{bmatrix} \begin{bmatrix} X(0) \\ \vdots \\ X(N-1) \end{bmatrix} + \begin{bmatrix} W(0) \\ \vdots \\ W(N-1) \end{bmatrix} \tag{7.13}
$$

The performance of the above system will be analyzed by deriving the BEP. The receiver is assumed to have an estimate of the channel at each subcarrier, namely, $\alpha(k)$, but this estimate is not necessarily accurate. Obtaining good channel estimates for a time-varying channel is a complicated task, and the quality of the channel estimates will determine the overall performance of the OFDM system. The process of obtaining the channel estimates is out of the scope of this chapter. For the interested reader, Refs. 3–6 provide a good sample of the work done on channel estimation for OFDM system using pilot symbol-aided modulation (PSAM). Since this chapter is mainly concerned with the overall performance analysis, the quality of the channel estimates will be measured by the complex correlation coefficient ρ between the channel $\alpha(k)$ and its estimate $g(k)$.

$$
\rho = \frac{E[\alpha g^*]}{\sqrt{E[|g|^2] E[|\alpha|^2]}} = \frac{Re\{E[\alpha g^*]\} + jIm\{E[\alpha g^*]\}}{\sigma_\alpha \sigma_g} \tag{7.14}
$$

where we have dropped the subcarrier index for simplicity of notation. $Re\{.\}$ and $Im\{.\}$ denote the real and imaginary parts, respectively, $2\sigma_g^2 = E[|g|^2]$ is the variance of the channel estimate and $E[.]$ the expectation operation. To show how ρ can be obtained, two common channel estimation schemes are considered. First, if the channel estimation scheme results in an additive error $(g = \alpha + \eta)$, where η is a complex Gaussian random variable with zero mean and variance $2\sigma_\eta^2$, then ρ is given by Ref. 15

$$
\rho = \frac{\sigma_\alpha}{\sqrt{\sigma_\alpha^2 + \sigma_\eta^2}}
$$

Second, if g is obtained from minimum mean square error channel estimation, then ρ is given by

$$
\rho = \frac{\sigma_g}{\sigma_\alpha}
$$

Starting from the received signal at the k_{th} subcarrier (Equation 7.5), after dropping the subcarrier index for notational simplicity

$$
Y = \alpha X + \beta + W
$$

The ICI is modeled as a zero-mean Gaussian random variable with variance equal to σ_β^2. A proof of the Gaussianity of the ICI is available in Ref. 7. Assuming M-QAM modulation, the equalized OFDM symbol is given by

$$Z = \frac{Y g^*}{|g|^2} = \frac{\alpha g^* X}{|g|^2} + \frac{(\beta + W)g^*}{|g|^2} \tag{7.15}$$

The direct approach to find the average BEP is to solve the following triple integration:

$$P_e = \int_0^\infty \int_0^\infty \int_0^{2\pi} P(e|\ |\alpha|, |g|, \phi) f(|\alpha|, |g|, \phi) \, d\alpha \, dg \, d\phi$$

where $P(e|\ |\alpha|, |g|, \phi)$ is the BEP conditioned on $|\alpha|, |g|$ and the phase difference between α and g, namely, ϕ, and $f(|\alpha|, |g|, \phi)$ is the joint probability density function (pdf) of $|\alpha|, |g|$, and ϕ. To avoid having to solve a triple integration, a different approach is proposed in Ref. 7 to simplify the problem.

Since α and g are jointly Gaussian [8], conditioned on g, the true channel α can be expressed in terms of the channel estimate g as follows [15]:

$$\alpha = \rho \frac{\sigma_\alpha}{\sigma_g} g + U \tag{7.16}$$

where U is a complex Gaussian random variable with independent inphase and quadrature Gaussian components. The random variable U has zero mean and variance equal to $\sigma_U^2 = 2\sigma_\alpha^2(1 - |\rho|^2)$. The variance of U represents the variance of the channel estimation error, while ρ represents the quality of the estimation. Substituting Equation 7.16 into Equation 7.15, we get

$$Z = \rho \frac{\sigma_\alpha}{\sigma_g} X + \frac{U g^* X}{|g|^2} + \frac{\beta g^*}{|g|^2} + \frac{W g^*}{|g|^2} \tag{7.17}$$

Now expanding U and W into inphase and quadrature component, the equalized OFDM symbol in Equation 7.17 becomes

$$Z = \rho \frac{\sigma_\alpha}{\sigma_g} X + \frac{\beta g^*}{|g|^2} + \frac{g^*(X U_I + W_I)}{|g|^2} + j \frac{g^*(X U_Q + W_Q)}{|g|^2}$$

$$= \rho \frac{\sigma_\alpha}{\sigma_g} X + \xi_1 + \xi_2 + j\xi_3 \tag{7.18}$$

Conditioned on g, ξ_1, ξ_2, and ξ_3 are Gaussian random variables with zero mean and variances given by

$$\sigma_{\xi_1}^2 = \frac{\sigma_\beta^2}{|g|^2}$$

$$\sigma_{\xi_{2,3}}^2 = \frac{\sigma_\alpha^2(1 - |\rho|^2) + \sigma_W^2}{|g|^2}$$

where it is assumed that $E[|X|^2]=1$. From the expression of the equalized OFDM symbol in Equation 7.18, it can been seen that the problem of finding the average error probability is greatly simplified, and reduced to solving a single integral to average over $|g|^2$ as opposed to triple integrations. To show the simplicity of the approach, the BEP for 16-QAM is derived. Figure 1 of Ref. 9 shows the 16-QAM constellation with gray coding. Assuming that each symbol contains four bits $(b_1b_2b_3b_4)$, the BEP for bit b_1, conditioned on g, is given by

$$P_e(b_1|g) = \frac{1}{2}[Pr(Z_I < 0|X_I = d, g) + Pr(Z_I < 0|X_I = 3d, g)] \tag{7.19}$$

where Z_I and X_I are the real parts of Z and X, respectively, and d is the minimum distance between two QAM symbols in the constellation. Similarly, the conditional error probability for bit b_3 is

$$P_e(b_3|g) = \frac{1}{2}[Pr(|Z_I| > 2d|X_I = d, g) + Pr(|Z_I| < 2d|X_I = 3d, g)] \tag{7.20}$$

Define $v = \xi_1 + \xi_2$, v is then a zero-mean Gaussian random variable with variance equal to $\sigma_v^2 = \sigma_{\xi_1}^2 + \sigma_{\xi_1}^2$, where it is assumed that the thermal noise is independent of the ICI. Then

$$P_e(b_1|g) = \frac{1}{2}\left[Q\left(\sqrt{\frac{\rho^2(\sigma_\alpha^2/\sigma_g^2)d^2}{\sigma_v^2}} \right) + Q\left(\sqrt{\frac{9\rho^2(\sigma_\alpha^2/\sigma_g^2)d^2}{\sigma_v^2}} \right) \right] \tag{7.21}$$

Substituting $\sigma_{\xi_1}^2$ and $\sigma_{\xi_2}^2$ into σ_v^2, the conditional BEP for bit b_1 becomes

$$P_e(b_1|g) = \frac{1}{2}\left[Q\left(\sqrt{\frac{\rho^2(\sigma_\alpha^2/\sigma_g^2)d^2|g|^2}{\sigma_\beta^2 + \sigma_\alpha^2(1-\rho^2) + \sigma_W^2}} \right) + Q\left(\sqrt{\frac{9\rho^2(\sigma_\alpha^2/\sigma_g^2)d^2|g|^2}{\sigma_\beta^2 + \sigma_\alpha^2(1-\rho^2) + \sigma_W^2}} \right) \right]$$
$$\tag{7.22}$$

Assuming a Rayleigh fading channel, the pdf of $|g|^2$ is $f_{|g|^2}(\lambda) = 1/2\sigma_g^2\, e^{-\lambda/2\sigma_g^2}$, and the average BER for bit b_1 is

$$P_e(b_1) = \frac{1}{2}(I_1 + I_2) \tag{7.23}$$

where

$$I_1 = \int_0^\infty Q\left(a_1\sqrt{|g|^2}\right) \frac{1}{2\sigma_g^2}e^{-\frac{\lambda}{2\sigma_g^2}}\, d\lambda$$

$$I_2 = \int_0^\infty Q\left(a_2\sqrt{|g|^2}\right) \frac{1}{2\sigma_g^2}e^{-\frac{\lambda}{2\sigma_g^2}}\, d\lambda$$

and

$$a_1^2 = \frac{\rho^2(\sigma_\alpha^2/\sigma_g^2)d^2}{\sigma_\beta^2 + \sigma_\alpha^2(1 - \rho^2) + \sigma_W^2}, \quad a_2^2 = 9a_1^2$$

By using Equation 5.6 in Ref. 10 we get

$$I_1 = \frac{1}{2}\left[1 - \sqrt{\frac{\rho^2\sigma_\alpha^2 d^2}{\sigma_\beta^2 + \sigma_\alpha^2(1 - \rho^2) + \sigma_W^2 + \rho^2\sigma_\alpha^2 d^2}}\right]$$

$$I_2 = \frac{1}{2}\left[1 - \sqrt{\frac{9\rho^2\sigma_\alpha^2 d^2}{\sigma_\beta^2 + \sigma_\alpha^2(1 - \rho^2) + \sigma_W^2 + 9\rho^2\sigma_\alpha^2 d^2}}\right]$$

then

$$P_e(b_1) = \frac{1}{2}\left[1 - \frac{1}{2}\sqrt{\frac{\rho^2\sigma_\alpha^2 d^2}{\sigma_\beta^2 + \sigma_\alpha^2(1 - \rho^2) + \sigma_W^2 + \rho^2\sigma_\alpha^2 d^2}}\right.$$
$$\left. - \frac{1}{2}\sqrt{\frac{9\rho^2\sigma_\alpha^2 d^2}{\sigma_\beta^2 + \sigma_\alpha^2(1 - \rho^2) + \sigma_W^2 + \rho^2\sigma_\alpha^2 d^2}}\right] \tag{7.24}$$

If we assume that the channel is time invariant ($\sigma_\beta^2 = 0$) and that perfect knowledge of the channel ($\rho = 1$) is available, then $P_e(b_1)$ reduces to the well-known result [11], namely,

$$P_e(b_1) = \frac{1}{2}\left[1 - \frac{1}{2}\sqrt{\frac{2\,snr}{5 + 2\,snr}} - \frac{1}{2}\sqrt{\frac{18\,snr}{5 + 18\,snr}}\right]$$

where $d^2 = 2E_b/5$ and $snr = \sigma_\alpha^2 E_b/\sigma_W^2$. Furthermore, if the channel is completely unknown ($\rho = 0$), then the average-error probability approaches 0.5, which is intuitively correct. By using similar steps as above, $P_e(b_3)$ is equal to

$$P_e(b_3) = \frac{1}{2}\left[1 - \frac{1}{2}\sqrt{\frac{d^2\eta_1^2\sigma_g^2}{\sigma_\beta^2 5\sigma_\alpha^2(1 - \rho^2) + \sigma_W^2 + d^2\eta_1^2\sigma_g^2}}\right.$$
$$- \frac{1}{2}\sqrt{\frac{d^2\eta_2^2\sigma_g^2}{\sigma_\beta^2 5\sigma_\alpha^2(1 - \rho^2) + \sigma_W^2 + d^2\eta_2^2\sigma_g^2}}$$
$$- \frac{1}{2}\sqrt{\frac{d^2\eta_3^2\sigma_g^2}{\sigma_\beta^2 5\sigma_\alpha^2(1 - \rho^2) + \sigma_W^2 + d^2\eta_3^2\sigma_g^2}}$$
$$\left. + \frac{1}{2}\sqrt{\frac{d^2\eta_4^2\sigma_g^2}{\sigma_\beta^2 5\sigma_\alpha^2(1 - \rho^2) + \sigma_W^2 + d^2\eta_4^2\sigma_g^2}}\right] \tag{7.25}$$

where $\eta_1 = 2 - \rho\,\sigma_\alpha/\sigma_g$, $\eta_2 = 2 + \rho\,\sigma_\alpha/\sigma_g$, $\eta_3 = -2 + 3\rho\,\sigma_\alpha/\sigma_g$, and $\eta_4 = 2 + 3\rho\,\sigma_\alpha/\sigma_g$. Owing to the symmetry of the 16-QAM constellation, it can be easily seen that $P_e(b_1) = P_e(b_2)$ and $P_e(b_3) = P_e(b_4)$. Hence, the final expression for the average BER is given by

$$P_e = \frac{1}{2}[P_e(b_1) + P_e(b_3)]$$

To complete the derivation, the power of the ICI, σ_β^2, needs to be derived. From Ref. 12,

$$\sigma_\beta^2 = E[|\beta(k)|^2]$$

$$= \frac{1}{N^2} \sum_{\substack{i=0 \\ i \neq k}}^{N-1} E[|X(i)|^2] \times \sum_{n_1=0}^{N-1}\sum_{n_2=0}^{N-1} E[H(n_1,i)H^*(n_2,i)]e^{\frac{-j2\pi(n_1-n_2)(k-i)}{N}}$$

$$= \frac{E_s}{N^2} \sum_{\substack{i=0 \\ i \neq k}}^{N-1} \left[N + 2\sum_{n=1}^{N-1}(N-n) \times J_0\left(\frac{2\pi f_d Tn}{N}\right)\cos\left(\frac{2\pi n(k-i)}{N}\right) \right]$$

The above procedure can be generalized to any constellation size and data modulation scheme. To investigate the performance of a specific channel estimation scheme, one has to only derive the appropriate value of ρ for that scheme, and express the true channel α in terms of its estimate g to simplify the problem.

7.3.2   Space-Time-Block-Coded (STBC) System

In Section 7.2.1, it is shown that the time variation of the wireless channel severely degrades the performance of OFDM; this degradation in performance is caused by the loss of orthogonality between subcarriers. In STBC-OFDM, the effect of wireless channel variation is more profound than that of SISO-OFDM, especially when orthogonal STBC are used to code across space and time. It will be shown that along with ICI, channel variations cause IAI between two different transmit antennas.

In the IEEE 802.16-2004 standard, Alamouti STBC [13] may be used on the downlink to provide second order (space) transmit diversity [1]. There are two transmit antennas on the base station (BS) side and one receive antenna on the subscriber station (SS) side. This scheme requires multiple-input-single-output channel estimation, which is out of the scope of this chapter. Figure 7.2 shows the STBC-OFDM as proposed by the standard documentation. Each transmit antenna has its own OFDM chain, and both antennas transmit at the same time two different OFDM symbols $x_1(n)$ and $x_2(n)$. The two symbols are transmitted twice from the two transmit antennas to get second-order diversity.

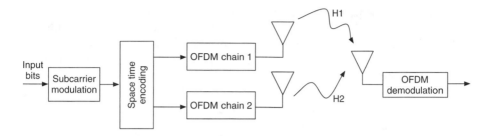

FIGURE 7.2
STBC-OFDM (Alamouti).

The transmission scheme can be described as follows: at time $2n$, OFDM symbols $x_1(n)$ and $x_2(n)$ are transmitted from the first and second antennas, respectively; at time $2n+1$, $-x_2^*(n)$ and $x_1^*(n)$ are transmitted from the first and second antennas, respectively, where $(.)^*$ denotes the conjugate operation. The code matrix of the above scheme is given by

$$\begin{bmatrix} x_1(n) & x_2(n) \\ -x_2^*(n) & x_1^*(n) \end{bmatrix} \tag{7.26}$$

Starting from the matrix representation of the received OFDM vector, and following the same notations in Equations 7.10 and 7.11, it can easily be seen that the received OFDM vectors for the $2n$ and $2n+1$ transmission instances are given by

$$\mathbf{Y}^{2n} = \mathbf{H}_1^{2n}\mathbf{X}_1 + \mathbf{H}_2^{2n}\mathbf{X}_2 + \mathbf{W}^{2n} \tag{7.27}$$

$$\mathbf{Y}^{2n+1} = -\mathbf{H}_1^{2n+1}\mathbf{X}_2^* + \mathbf{H}_2^{2n+1}\mathbf{X}_1^* + \mathbf{W}^{2n+1} \tag{7.28}$$

where $\mathbf{Y}^i$ and $\mathbf{W}^i$ are the received OFDM vector and AWGN at time $i = 2n, 2n+1$, respectively; $\mathbf{H}_j^i$ is the channel matrix from transmit antenna $j = 1, 2$ at time i and $\mathbf{X}_j$ the jth transmitted OFDM vectors. The above system can be written in the following compact matrix representation:

$$\begin{bmatrix} \mathbf{Y}^{2n} \\ \mathbf{Y}^{2n+1*} \end{bmatrix} = \begin{bmatrix} \mathbf{H}_1^{2n} & \mathbf{H}_2^{2n} \\ \mathbf{H}_2^{2n+1*} & -\mathbf{H}_1^{2n+1*} \end{bmatrix} \begin{bmatrix} \mathbf{X}_1 \\ \mathbf{X}_2 \end{bmatrix} + \begin{bmatrix} \mathbf{W}^{2n} \\ \mathbf{W}^{2n+1*} \end{bmatrix}$$

$$\mathbf{Y} = \mathbf{H}\mathbf{X} + \mathbf{W} \tag{7.29}$$

where $\mathbf{Y}, \mathbf{X}, \mathbf{W} \in C^{2N \times 1}$, and $\mathbf{H} \in C^{2N \times 2N}$. By carefully analyzing Equation 7.29, the following two problems can be observed if the channel is time varying.

- Similar to the SISO system in Equation 7.13, the channel matrices $\mathbf{H}_j^i$ are not diagonal.
- $\mathbf{H}_1^{2n} \neq \mathbf{H}_1^{2n+1}$ and $\mathbf{H}_2^{2n} \neq \mathbf{H}_2^{2n+1}$.

As a direct consequence of the first problem, the system performance will be degraded from ICI caused by the nondiagonal elements of $\mathbf{H}_j^i$. This ICI is because of channel variations and is independent of the coding scheme used. The second problem causes IAI between the two transmit antennas, which further degrades system performance. Analogous to ICI, the IAI results from the loss of orthogonality of the channel matrix, a major property of the Alamouti coding scheme; hence, IAI depends on the code used, and is present only in orthogonal STBC schemes.

The effect of the time variation of the wireless channel can also be seen by observing the received signal at a specific subcarrier. The received OFDM signal at the kth subcarrier is given by

$$Y^{2n} = \alpha_1^{2n} X_1 + \beta_1^{2n} + \alpha_2^{2n} X_2 + \beta_2^{2n} + W^{2n} \tag{7.30}$$

$$Y^{2n+1} = -\alpha_1^{2n+1} X_2^* - \beta_1^{2n+1} + \alpha_2^{2n+1} X_1^* + \beta_2^{2n+1} + W^{2n+1} \tag{7.31}$$

and in matrix form

$$\begin{bmatrix} Y^{2n} \\ Y^{2n+1*} \end{bmatrix} = \begin{bmatrix} \alpha_1^{2n} & \alpha_2^{2n} \\ \alpha_2^{2n+1*} & -\alpha_1^{2n+1*} \end{bmatrix} \begin{bmatrix} X_1 \\ X_2 \end{bmatrix} + \begin{bmatrix} \beta_1^{2n} + \beta_2^{2n} \\ \beta_2^{2n+1} - \beta_1^{2n+1} \end{bmatrix} + \begin{bmatrix} W^{2n} \\ W^{2n+1*} \end{bmatrix} \tag{7.32}$$

where the subcarrier index is dropped for notational simplicity. α_j^i and β_j^i are the same as those defined in Equation 7.5. The two problems discussed above can be clearly seen here and restated as such:

- Time variation of the wireless channel causes ICI.
- The channel matrix is not orthogonal anymore, because $\alpha_j^{2n} \neq \alpha_j^{2n+1}$; hence, IAI is present.

Figure 7.3 shows the block diagram of the proposed detector. The first step in decoding the transmitted symbols, as proposed by Alamouti, is to multiply the received OFDM vector in Equation 7.29 by the transpose conjugate of the channel matrix; this operation is referred to as space time combining.

$$\mathbf{Z} = \mathbf{H}^H \mathbf{Y} = \mathbf{H}^H (\mathbf{H}\mathbf{X} + \mathbf{W})$$

$$= \mathbf{H}^H \mathbf{H}\mathbf{X} + \mathbf{H}^H \mathbf{W}$$

$$= \mathbf{A}\mathbf{X} + \hat{\mathbf{W}} \tag{7.33}$$

$$\begin{bmatrix} \mathbf{Z}^{2n} \\ \mathbf{Z}^{2n+1} \end{bmatrix} = \begin{bmatrix} \mathbf{A}_1 & \mathbf{A}_2 \\ \mathbf{A}_3 & \mathbf{A}_4 \end{bmatrix} \begin{bmatrix} \mathbf{X}_1 \\ \mathbf{X}_2 \end{bmatrix} + \begin{bmatrix} \hat{\mathbf{W}}^{2n} \\ \hat{\mathbf{W}}^{2n+1} \end{bmatrix} \tag{7.34}$$

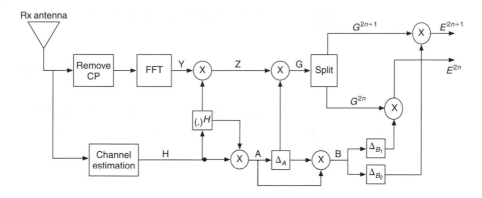

FIGURE 7.3
Block diagram of the proposed detection for STBC-OFDM.

where

$$\mathbf{A} = \mathbf{H}^H \mathbf{H}$$

$$\mathbf{A}_1 = \mathbf{H}_1^{2n^H} \mathbf{H}_1^{2n} + \mathbf{H}_2^{2n+1^T} \mathbf{H}_2^{2n+1^*}$$

$$\mathbf{A}_2 = \mathbf{H}_1^{2n^H} \mathbf{H}_2^{2n} - \mathbf{H}_2^{2n \mid 1^T} \mathbf{H}_1^{2n \mid 1^*}$$

$$\mathbf{A}_3 = \mathbf{H}_2^{2n^H} \mathbf{H}_1^{2n} - \mathbf{H}_1^{2n+1^T} \mathbf{H}_2^{2n+1^*}$$

$$\mathbf{A}_4 = \mathbf{H}_1^{2n^H} \mathbf{H}_1^{2n} + \mathbf{H}_2^{2n+1^T} \mathbf{H}_2^{2n+1^*}$$

and $\mathbf{A}_1$, $\mathbf{A}_2$, $\mathbf{A}_3$, and $\mathbf{A}_4$ are all $\in C^{N \times N}$. If the channel is time invariant, then $\mathbf{H}$ is an orthogonal matrix and $\mathbf{A}_2 = \mathbf{A}_3 = 0$; therefore, Equation 7.34 reduces to

$$\begin{bmatrix} \mathbf{Z}^{2n} \\ \mathbf{Z}^{2n+1} \end{bmatrix} = \begin{bmatrix} \mathbf{A}_1 & 0 \\ 0 & \mathbf{A}_4 \end{bmatrix} \begin{bmatrix} \mathbf{X}_1 \\ \mathbf{X}_2 \end{bmatrix} + \begin{bmatrix} \hat{\mathbf{W}}^{2n} \\ \hat{\mathbf{W}}^{2n+1} \end{bmatrix} \qquad (7.35)$$

where $\mathbf{A}_1$ and $\mathbf{A}_4$ are diagonal matrices, and 0 are matrices with zeros in all its elements. It can be seen from Equation 7.35 that $\mathbf{X}_1$ and $\mathbf{X}_2$ are completely decoupled, and can accurately be detected using the maximum likelihood (ML) detector proposed in Ref. 13.

In a mobile channel environment, the transmitted symbols $\mathbf{X}_1$ and $\mathbf{X}_2$ cannot be decoupled using the Alamouti detection scheme, and a different approach should be taken. Here, a new two-stage detection scheme that will reduce the effect of both ICI and IAI and enable us to accurately detect the transmitted symbols is presented.

Starting from the matrix representation of the output of the space time combiner

$$\begin{bmatrix} \mathbf{Z}^{2n} \\ \mathbf{Z}^{2n+1} \end{bmatrix} = \begin{bmatrix} \mathbf{A}_1 & \mathbf{A}_2 \\ \mathbf{A}_3 & \mathbf{A}_4 \end{bmatrix} \begin{bmatrix} \mathbf{X}_1 \\ \mathbf{X}_2 \end{bmatrix} + \begin{bmatrix} \hat{\mathbf{W}}^{2n} \\ \hat{\mathbf{W}}^{2n+1} \end{bmatrix}$$

$$\mathbf{Z} = \mathbf{A}\mathbf{X} + \hat{\mathbf{W}}$$

In the first stage of the detector, multiply the output of the space time combiner by the following decoupling matrix [14]:

$$\boldsymbol{\Delta}_A = \begin{bmatrix} \mathbf{I}_N & -\mathbf{A}_2\mathbf{A}_4^{-1} \\ -\mathbf{A}_3\mathbf{A}_1^{-1} & \mathbf{I}_N \end{bmatrix} \tag{7.36}$$

where $\boldsymbol{\Delta}_A$ is a decoupling matrix constructed from the elements of $\mathbf{A}$, and $\mathbf{I}_N$ is an $N \times N$ identity matrix. The result of the first stage would be

$$\mathbf{G} = \boldsymbol{\Delta}_A\mathbf{Z} = \boldsymbol{\Delta}_A\mathbf{A}\mathbf{X} + \boldsymbol{\Delta}_A\hat{\mathbf{W}}$$
$$= \mathbf{B}\mathbf{X} + \tilde{\mathbf{W}} \tag{7.37}$$

where the matrices in Equation 7.37 are defined as follows:

$$\mathbf{G} = \boldsymbol{\Delta}_A\,\mathbf{Z} = \begin{bmatrix} \mathbf{G}^{2n} \\ \mathbf{G}^{2n+1} \end{bmatrix} \quad \in C^{2N \times 1} \tag{7.38}$$

$$\mathbf{B} = \boldsymbol{\Delta}_A\,\mathbf{A} = \begin{bmatrix} \mathbf{B}_1 & 0 \\ 0 & \mathbf{B}_2 \end{bmatrix} \quad \in C^{2N \times 2N} \tag{7.39}$$

$$\mathbf{B}_1 = \mathbf{A}_1 - \mathbf{A}_2\mathbf{A}_4^{-1}\mathbf{A}_3 \quad \in C^{N \times N} \tag{7.40}$$

$$\mathbf{B}_2 = \mathbf{A}_4 - \mathbf{A}_3\mathbf{A}_1^{-1}\mathbf{A}_2 \quad \in C^{N \times N} \tag{7.41}$$

and the output of the first stage is

$$\begin{bmatrix} \mathbf{G}^{2n} \\ \mathbf{G}^{2n+1} \end{bmatrix} = \begin{bmatrix} \mathbf{B}_1 & 0 \\ 0 & \mathbf{B}_2 \end{bmatrix}\begin{bmatrix} \mathbf{X}_1 \\ \mathbf{X}_2 \end{bmatrix} + \begin{bmatrix} \tilde{\mathbf{W}}^{2n} \\ \tilde{\mathbf{W}}^{2n+1} \end{bmatrix} \tag{7.42}$$

It can be seen from Equation 7.42 that the transmitted symbols $\mathbf{X}_1$ and $\mathbf{X}_2$ are completely decoupled, but the effect of ICI is still present because $\mathbf{B}_1$ and $\mathbf{B}_2$ are not diagonal. $\mathbf{B}_1$ and $\mathbf{B}_2$ have the following structure:

$$\mathbf{B}_1 = \begin{bmatrix} b_{0,0}^1 & b_{0,1}^1 & \cdots & b_{0,N-1}^1 \\ b_{1,0}^1 & b_{1,1}^1 & \cdots & b_{1,N-1}^1 \\ \vdots & \vdots & \ddots & \vdots \\ b_{N-1,0}^1 & b_{N-1,1}^1 & \cdots & b_{N-1,N-1}^1 \end{bmatrix}$$

$$\mathbf{B}_2 = \begin{bmatrix} b_{0,0}^2 & b_{0,1}^2 & \cdots & b_{0,N-1}^2 \\ b_{1,0}^2 & b_{1,1}^2 & \cdots & b_{1,N-1}^2 \\ \vdots & \vdots & \ddots & \vdots \\ b_{N-1,0}^2 & b_{N-1,1}^2 & \cdots & b_{N-1,N-1}^2 \end{bmatrix}$$

where $b^j_{p,q}$ is the (p, q) element of $\mathbf{B}_j$. Similar to what was discussed earlier, the nondiagonal elements of $\mathbf{B}_1$ and $\mathbf{B}_2$ are the ICI caused by channel variations. The goal of the second stage of the detector is to reduce the effect of ICI.

In the second-detection stage, split the output of the first stage into even and odd subcarriers such as

$$\mathbf{G}^{2n} = \begin{bmatrix} \mathbf{G}^{2n}_e \\ \mathbf{G}^{2n}_o \end{bmatrix} = \begin{bmatrix} \mathbf{B}_{11} & \mathbf{B}_{12} \\ \mathbf{B}_{13} & \mathbf{B}_{14} \end{bmatrix} \begin{bmatrix} \mathbf{X}_{1e} \\ \mathbf{X}_{1o} \end{bmatrix} + \begin{bmatrix} \tilde{\mathbf{W}}^{2n}_e \\ \tilde{\mathbf{W}}^{2n}_o \end{bmatrix} \tag{7.43}$$

$$\mathbf{G}^{2n+1} = \begin{bmatrix} \mathbf{G}^{2n+1}_e \\ \mathbf{G}^{2n+1}_o \end{bmatrix} = \begin{bmatrix} \mathbf{B}_{21} & \mathbf{B}_{22} \\ \mathbf{B}_{23} & \mathbf{B}_{24} \end{bmatrix} \begin{bmatrix} \mathbf{X}_{2e} \\ \mathbf{X}_{2o} \end{bmatrix} + \begin{bmatrix} \tilde{\mathbf{W}}^{2n+1}_e \\ \tilde{\mathbf{W}}^{2n+1}_o \end{bmatrix} \tag{7.44}$$

where

$$\mathbf{G}^i_e = \begin{bmatrix} G^i(0) & G^i(2) & \cdots & G^i(N-2) \end{bmatrix}^{\mathrm{T}} \in C^{\frac{N}{2} \times 1}, \ i = 2n, 2n+1 \tag{7.45}$$

$$\mathbf{G}^i_o = \begin{bmatrix} G^i(1) & G^i(3) & \cdots & G^i(N-1) \end{bmatrix}^{\mathrm{T}} \in C^{\frac{N}{2} \times 1}, \ i = 2n, 2n+1 \tag{7.46}$$

$$\mathbf{X}_{je} = \begin{bmatrix} X_j(0) & X_j(2) & \cdots & X_j(N-2) \end{bmatrix}^{\mathrm{T}} \in C^{\frac{N}{2} \times 1}, \ j = 1, 2 \tag{7.47}$$

$$\mathbf{X}_{jo} = \begin{bmatrix} X_j(1) & X_i(3) & \cdots & X_i(N-1) \end{bmatrix}^{\mathrm{T}} \in C^{\frac{N}{?} \times 1}, \ j = 1, 2 \tag{7.48}$$

and

$$\mathbf{B}_{k1} = \begin{bmatrix} b^k_{0,0} & b^k_{0,2} & \cdots & b^k_{0,N-2} \\ b^k_{2,0} & b^k_{2,2} & \cdots & b^1_{2,N-2} \\ \vdots & \vdots & \ddots & \vdots \\ b^k_{N-2,0} & b^k_{N-2,2} & \cdots & b^k_{N-2,N-2} \end{bmatrix} \in C^{\frac{N}{2} \times \frac{N}{2}}, \ k = 1, 2 \tag{7.49}$$

$$\mathbf{B}_{k2} = \begin{bmatrix} b^k_{0,1} & b^k_{0,3} & \cdots & b^k_{0,N-1} \\ b^k_{2,1} & b^k_{2,3} & \cdots & b^k_{2,N-1} \\ \vdots & \vdots & \ddots & \vdots \\ b^k_{N-2,1} & b^k_{N-2,3} & \cdots & b^k_{N-2,N-1} \end{bmatrix} \in C^{\frac{N}{2} \times \frac{N}{2}}, \ k = 1, 2 \tag{7.50}$$

$$\mathbf{B}_{k3} = \begin{bmatrix} b^k_{1,0} & b^k_{1,2} & \cdots & b^k_{1,N-2} \\ b^k_{3,0} & b^k_{3,2} & \cdots & b^k_{3,N-2} \\ \vdots & \vdots & \ddots & \vdots \\ b^k_{N-1,0} & b^k_{N-1,2} & \cdots & b^k_{N-1,N-2} \end{bmatrix} \in C^{\frac{N}{2} \times \frac{N}{2}}, \ k = 1, 2 \tag{7.51}$$

$$\mathbf{B}_{k4} = \begin{bmatrix} b_{1,1}^k & b_{1,3}^k & \cdots & b_{1,N-1}^k \\ b_{3,1}^k & b_{3,3}^k & \cdots & b_{3,N-1}^k \\ \vdots & \vdots & \ddots & \vdots \\ b_{N-1,1}^k & b_{N-1,3}^k & \cdots & b_{N-1,N-1}^k \end{bmatrix} \in C^{\frac{N}{2} \times \frac{N}{2}}, \ k = 1, 2 \qquad (7.52)$$

An interesting result from the above operation is that the matrices $\mathbf{B}_{k1}$ and $\mathbf{B}_{k4}$ contain partial ICI, in the sense that they have ICI contribution from half of the subcarriers only, and $\mathbf{B}_{k2}$ and $\mathbf{B}_{k3}$ are full ICI matrices. It is desired now to eliminate $\mathbf{B}_{k2}$ and $\mathbf{B}_{k3}$. To achieve this goal, multiply $\mathbf{G}^{2n}$ and $\mathbf{G}^{2n+1}$ in Equations 7.43 and 7.44 with the following matrices respectively.

$$\mathbf{\Delta}_{B_1} = \begin{bmatrix} \mathbf{I}_{\frac{N}{2}} & -\mathbf{B}_{12}\mathbf{B}_{14}^{-1} \\ -\mathbf{B}_{13}\mathbf{B}_{11}^{-1} & \mathbf{I}_{\frac{N}{2}} \end{bmatrix} \qquad (7.53)$$

$$\mathbf{\Delta}_{B_2} = \begin{bmatrix} \mathbf{I}_{\frac{N}{2}} & -\mathbf{B}_{22}\mathbf{B}_{24}^{-1} \\ -\mathbf{B}_{23}\mathbf{B}_{21}^{-1} & \mathbf{I}_{\frac{N}{2}} \end{bmatrix} \qquad (7.54)$$

That is,

$$\mathbf{E}^{2n} = \begin{bmatrix} \mathbf{E}_e^{2n} \\ \mathbf{E}_o^{2n} \end{bmatrix} = \mathbf{\Delta}_{B_1}\mathbf{G}^{2n} = \mathbf{\Delta}_{B_1} \begin{bmatrix} \mathbf{B}_{11} & \mathbf{B}_{12} \\ \mathbf{B}_{13} & \mathbf{B}_{14} \end{bmatrix} \begin{bmatrix} \mathbf{X}_{1e} \\ \mathbf{X}_{1o} \end{bmatrix} + \mathbf{\Delta}_{B_1} \begin{bmatrix} \tilde{\mathbf{W}}_e^{2n} \\ \tilde{\mathbf{W}}_o^{2n} \end{bmatrix} \qquad (7.55)$$

$$\mathbf{E}^{2n+1} = \begin{bmatrix} \mathbf{E}_e^{2n+1} \\ \mathbf{E}_o^{2n+1} \end{bmatrix} = \mathbf{\Delta}_{B_2}\mathbf{G}^{2n+1} = \mathbf{\Delta}_{B_2} \begin{bmatrix} \mathbf{B}_{21} & \mathbf{B}_{22} \\ \mathbf{B}_{23} & \mathbf{B}_{24} \end{bmatrix} \begin{bmatrix} \mathbf{X}_{2e} \\ \mathbf{X}_{2o} \end{bmatrix} + \mathbf{\Delta}_{B_2} \begin{bmatrix} \tilde{\mathbf{W}}_e^{2n+1} \\ \tilde{\mathbf{W}}_o^{2n+1} \end{bmatrix}$$

$$(7.56)$$

and the result is

$$\begin{bmatrix} \mathbf{E}_e^{2n} \\ \mathbf{E}_o^{2n} \end{bmatrix} = \begin{bmatrix} \mathbf{\Lambda}_1 & 0 \\ 0 & \mathbf{\Lambda}_2 \end{bmatrix} \begin{bmatrix} \mathbf{X}_{1e} \\ \mathbf{X}_{1o} \end{bmatrix} + \begin{bmatrix} \bar{\mathbf{W}}_e^{2n} \\ \bar{\mathbf{W}}_o^{2n} \end{bmatrix} \qquad (7.57)$$

$$\begin{bmatrix} \mathbf{E}_e^{2n+1} \\ \mathbf{E}_o^{2n+1} \end{bmatrix} = \begin{bmatrix} \mathbf{\Lambda}_3 & 0 \\ 0 & \mathbf{\Lambda}_4 \end{bmatrix} \begin{bmatrix} \mathbf{X}_{2e} \\ \mathbf{X}_{2o} \end{bmatrix} + \begin{bmatrix} \bar{\mathbf{W}}_e^{2n+1} \\ \bar{\mathbf{W}}_o^{2n+1} \end{bmatrix} \qquad (7.58)$$

where

$$\mathbf{\Lambda}_1 = \mathbf{B}_{11} - \mathbf{B}_{12}\mathbf{B}_{14}^{-1}\mathbf{B}_{13} \quad \in C^{\frac{N}{2} \times \frac{N}{2}} \qquad (7.59)$$

$$\mathbf{\Lambda}_2 = \mathbf{B}_{14} - \mathbf{B}_{13}\mathbf{B}_{11}^{-1}\mathbf{B}_{12} \quad \in C^{\frac{N}{2} \times \frac{N}{2}} \qquad (7.60)$$

$$\boldsymbol{\Lambda}_3 = \mathbf{B}_{21} - \mathbf{B}_{22}\mathbf{B}_{24}^{-1}\mathbf{B}_{23} \quad \in C^{\frac{N}{2}\times\frac{N}{2}} \tag{7.61}$$

$$\boldsymbol{\Lambda}_4 = \mathbf{B}_{24} - \mathbf{B}_{23}\mathbf{B}_{21}^{-1}\mathbf{B}_{22} \quad \in C^{\frac{N}{2}\times\frac{N}{2}} \tag{7.62}$$

and the final expression would be

$$\mathbf{E}_e^{2n} = \boldsymbol{\Lambda}_1\,\mathbf{X}_{1e} + \bar{\mathbf{W}}_e^{2n} \tag{7.63}$$

$$\mathbf{E}_o^{2n} = \boldsymbol{\Lambda}_2\,\mathbf{X}_{1o} + \bar{\mathbf{W}}_o^{2n} \tag{7.64}$$

$$\mathbf{E}_e^{2n+1} = \boldsymbol{\Lambda}_3\,\mathbf{X}_{2e} + \bar{\mathbf{W}}_e^{2n+1} \tag{7.65}$$

$$\mathbf{E}_o^{2n+1} = \boldsymbol{\Lambda}_4\,\mathbf{X}_{2o} + \bar{\mathbf{W}}_o^{2n+1} \tag{7.66}$$

After the second stage, the two objectives have been achieved, namely, successful decoupling of the transmitted symbols and reduction of ICI. The transmitted symbols can now be detected by using the ML detector proposed by Alamouti [13]. The average BEP expression for the above system can be derived by first writing Equations 7.63 through 7.66 in a scalar form. Next, the power of the ICI is derived using an approach similar to the one used in deriving σ_β^2 in Section 7.3.1. Finally, using Equations 7.15 through 7.25, the expression for the average BEP can be derived. The actual expressions are omitted from this chapter because of space limitations.

7.4 Numerical and Simulation Results

This section presents some simulation results of BEP for the two systems considered earlier in this chapter. First, the BEP of SISO-OFDM in the case of perfect channel estimation is considered. Figures 7.4 and 7.5 show the average BEP of 16-QAM OFDM with 256 subcarriers for different mobile speeds with 20 and 1.75 MHz channel bandwidth, respectively. Clearly, the average BEP suffer from error floor owing to channel variations. The error floor is more severe in Figure 7.5 because the symbol duration is larger and the system is more susceptible to channel variations. Figure 7.6 compares the BEP at 40 dB E_b/N_0 plotted against different mobile speeds for a 16-QAM OFDM system with 256 subcarriers and different symbol duration.

The average BEP performance of SISO-OFDM with channel estimation errors is shown in Figures 7.7 and 7.8. Figure 7.7 shows the average BEP performance of 16-QAM OFDM with 256 subcarriers, 20 MHz channel bandwidth, 3 km/h mobile speed, and different values of the complex correlation coefficient ρ. Figure 7.8 shows the average BEP performance at 100 km/h mobile speed, using the same system parameters used in Figure 7.7.

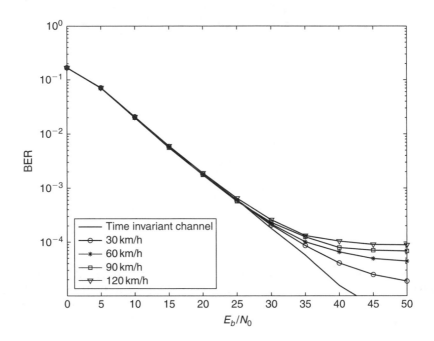

FIGURE 7.4
Average BEP for 16-QAM SISO-OFDM with 256 subcarriers and 11.2 μs symbol duration for different mobile speeds with perfect channel estimation.

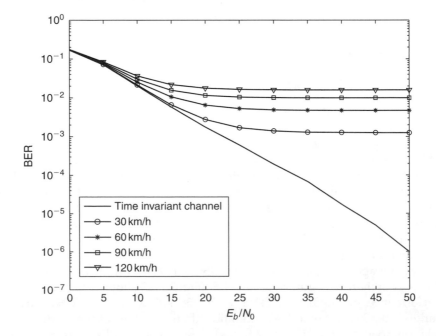

FIGURE 7.5
Average BEP for 16-QAM SISO-OFDM with 256 subcarriers and 148 μs symbol duration for different mobile speeds with perfect channel estimation.

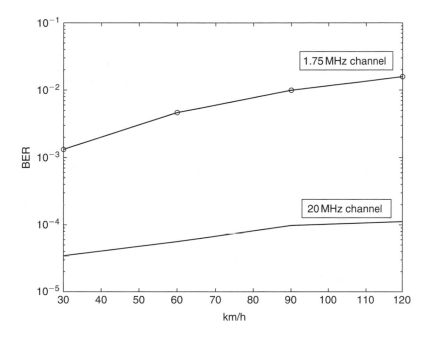

FIGURE 7.6
Average BEP comparison at 40 E_b/N_0 with perfect channel estimation.

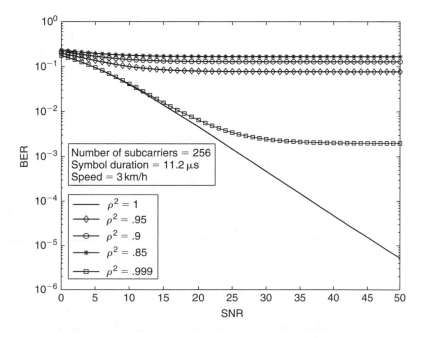

FIGURE 7.7
Average BEP for 16-QAM SISO-OFDM with 256 subcarriers and 11.2 μs symbol duration at 3 km/h with channel estimation errors.

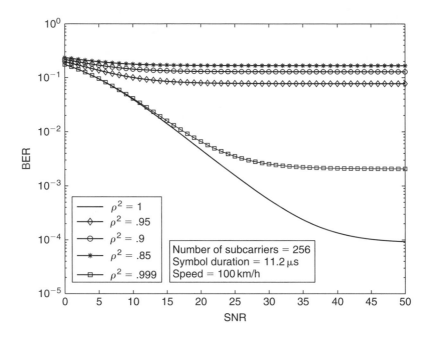

FIGURE 7.8
Average BEP for 16-QAM SISO-OFDM with 256 subcarriers and 11.2 μs symbol duration at 100 km/h with channel estimation errors.

The combined effect of channel variation and channel estimation errors can be clearly seen from the large error floor. Also, with channel estimation errors, the error floor occurs at low values of E_b/N_0.

Next, the performance of STBC-OFDM is considered. Figures 7.9 through 7.11 compare the average BEP performance of STBC-OFDM with and without IAI cancellation and ICI reduction. In Figure 7.9, the average BEP performance of 16-QAM STBC-OFDM with 256 subcarriers and channel bandwidth of 20 MHz is plotted versus E_b/N_0. It can be seen that the standard detection will cause a high-error floor, especially at high mobile speeds. The improvement in BEP when employing the proposed detection can also be seen. Figures 7.10 and 7.11 show the average BEP for 1.75 MHz channel bandwidth. Despite the significant improvement in average BEP, the system still suffers from high-error floor compared to the 20 MHz channel bandwidth case. The reason is that, although the number of interfering subcarriers is reduced by half when using the proposed detector, the power of the ICI from the first and second adjacent subcarriers is still high. One way to improve the performance is to repeat the second stage of the proposed detector. This will further eliminate some of the high-power interfering subcarriers, and lower the error floor. A trade-off between performance and complexity should determine the number of times the second stage is repeated.

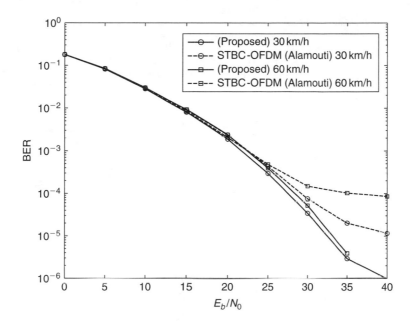

FIGURE 7.9
Average BEP comparison between the standard Alamouti and the proposed detection schemes for 16-QAM MISO-OFDM with 256 subcarriers and 11.2 μs symbol duration at 30 and 60 km/h.

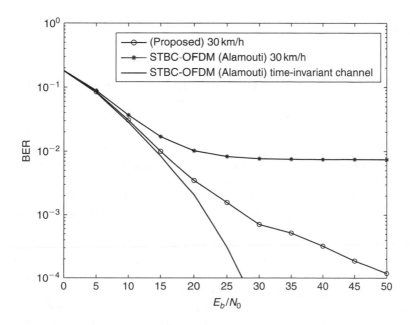

FIGURE 7.10
Average BEP comparison between the standard Alamouti and the proposed detection schemes for 16-QAM MISO-OFDM with 256 subcarriers and 148 μs symbol duration at 30 km/h.

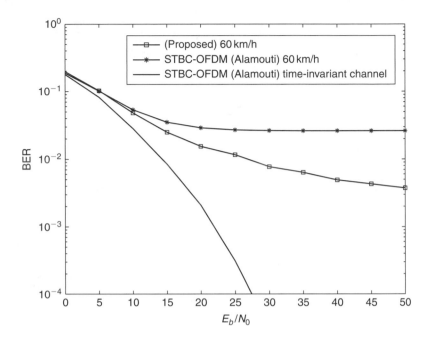

FIGURE 7.11
Average BEP comparison between the standard Alamouti and the proposed detection schemes for 16-QAM MISO-OFDM with 256 subcarriers and 148 μs symbol duration at 60 km/h.

7.5 Conclusion

In this chapter, the performance of both SISO- and STBC-OFDM systems are analyzed in mobile environments. In the SISO-OFDM case, the average BEP is derived assuming the general case when the receiver does not have perfect knowledge of the channel state information. It was shown that by conveniently expressing the actual channel in terms of the channel estimates, the BEP expressions can simply be derived by solving a single integral as opposed to the threefold integral suggested in prevues literature. For the STBC-OFDM, it was shown that if the channel is time varying, IAI will also be present in addition to ICI, and the conventional Alamouti-detection scheme will fail to detect the transmitted OFDM signal. To overcome this problem, a two-stage detection scheme was suggested. The first stage of the new detector eliminates IAI, where the second stage reduces ICI. To further improve the BEP performance, the second step can be repeated many times to reach the desired performance level. A trade-off between performance and complexity should determine the number of times the second stage is repeated. Simulation results show the advantages of employing the new detection scheme as opposed to the conventional Alamouti scheme. All simulations are performed using the same parameters of the IEEE 802.16-2004 standard.

References

1. *IEEE Standard for Local and Metropolitan Area Networks, Part 16: Air Interface for Fixed Broadband Wireless Access Systems*, IEEE 2004.
2. V. Erceg, K.V.S. Hari, M.S. Smith, D.S. Baum, K.P. Sheikh, C. Tappenden, J.M. Costa, C. Bashue, A. Sarajedini, R. Shwartz, and D. Branlund, *Channel Models for Fixed Wireless Applications*, Contribution IEEE 802.16.3c-01/29rl February 2001.
3. S. Coleri, M. Ergen, A. Puri, and A. Bahai, Channel estimation techniques based on pilot arrangement in OFDM systems, *IEEE Trans. Broad.*, vol. 48, No. 3, pp. 223–229, September 2002.
4. S.G. Kang, Y.M. Ha, and E.K. Joo, A comparative investigation on channel estimation algorithms for OFDM in mobile communications, *IEEE Trans. Broad.*, vol. 49, No. 2, pp. 142–154, June 2003.
5. O. Edford, et al., OFDM channel estimation by singular value decomposition, *IEEE Trans. Commun.*, vol. 46, No. 7, pp. 931–939, July 1998.
6. M. Morelli and U. Mengali, A comparison of pilot-aided channel estimation methods for ofdm systems, *IEEE Trans. Sig. Proc.*, vol. 49, No. 12, pp. 3065–3073, December 2001.
7. M. Al-Gharabally and P. Das, On the performance of OFDM systems in time varying channels with channel estimation error. In Proc. *IEEE ICC-2006*, vol. 11, pp. 5180–5158, June 2006.
8. Z.-Z. Chang and Y.T. Su, Performance analysis of equalized ofdm systems in rayleigh fading, *IEEE Trans. Wireless Comm.*, vol. 1, No. 4, pp. 721–732, October 2002.
9. L. Cao and N.C. Beaulieu, Exact error-rate analysis of diversity 16-qam with channel estimation error, *IEEE Trans. Commun.*, vol. 52, No. 6, pp. 1019–1029, June 2004.
10. M.K. Simon and M.S. Alouini, *Digital Communications over Generalized Fading Channels: A Unified Approach to Performance Analysis*, Wiley and Sons, New York, 2000.
11. M. Surendra Raju, A. Ramesh, and A. Chockalingam, BER analysis of QAM with transmit diversity in Rayleigh fading channels, *IEEE GLOBECOM 2003*, vol. 2, pp. 641–645, December 2003.
12. Y.-S. Choi, P.J. Voltz, and F.A. Cassara, On channel estimation and detection for multicarrier signals in fast and selective rayleigh fading channels, *IEEE Trans. Commun.*, vol. 49, No. 8, pp. 1375–1376, August 2001.
13. S.M. Alamouti, A simple transmitter diversity scheme for wireless communications, *IEEE J. Select. Areas Comm.*, vol. 16, pp. 1451–1458, October 1998.
14. W.M. Younis, A.H. Sayed, and N. Al-Dhahir, Efficient adaptive receivers for joint equalization and interface cancellation in multiuser space-time block-coded systems, *IEEE Trans. Sig. Proc.*, vol. 51, No. 11, pp. 2849–2862, November 2003.
15. R. Annavajjala, P.C. Cosman, and L.B. Milstein, Performance analysis of linear modulation schemes with generalized diversity combining on Rayleigh fading channels with noisy channel estimates, *IEEE Trans. Info. Theory*, under revision.

Part III

QoS

8

IEEE 802.16 Multiple Access Control:
Resources Allocation for Reservation-Based
Traffic

Ahmed Doha and Hossam Hassanein

CONTENTS

The IEEE 802.16 standard defines the physical layer (PHY) and multiple access control (MAC) layer specifications of wireless metropolitan area networks (WirelessMAN). The advent of the IEEE 802.16 standard lays technological grounds for the support of important and massively demanded wireless broadband applications that interconnect homes and businesses. With relatively extended radio frequency (RF) coverage, the IEEE 802.16 networks can be a technologically and economically attractive alternative to the traditional cable networks including digital subscriber line xDSL, fiber optics, and coaxial cable networks. Technological sophistication is intrinsic to the design of the uplink and downlink of the IEEE 802.16 MAC, offering higher data rates and supporting hundreds of data terminals per link that can exchange various types of traffic with various quality of service (QoS) requirements. However, avoiding the construction work and civil permits associated with laying underground cable networks distinguishes the IEEE 802.16 networks both from a service-delivery schedule and implementation economics perspectives.

8.1 Introduction

The IEEE 802.16 standard is emerging as a QoS-rich platform. Different access methods are supported for different classes of traffic. Best effort (BE) traffic is one of the most important of these classes as it represents the majority of the overall data traffic. The standard specified that the MAC protocol should use a reservation-based access method for the BE traffic. However, it does not recommend a specific reservation-based MAC (R-MAC) protocol to be used, an area that was left for product differentiation. Although R-MAC protocols have received considerable attention in the literature, the imminent proliferation of the IEEE 802.16 standard reenergized a need to closely understand and enhance their performance.

R-MAC protocols have been a primary access method for broadband access technologies. For example, the general packet radio system (GPRS), digital subscriber line (DSL), and hybrid fiber coaxial (HFC) cable technologies employ reservation-based multiple access systems. Nevertheless, IEEE 802 WirelessMAN also employs an R-MAC technology.

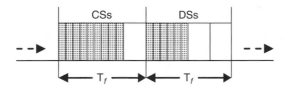

FIGURE 8.1
Example frame structure of R-MAC system.

In R-MAC protocols, time is divided into frames where the physical slots of each frame can be sued for contention-based reservation or data transmission [1] as shown in Figure 8.1. In most of the commercial versions of R-MAC, the service period is managed using time division multiple access (TDMA) owing to its simplicity. The reservation mechanism is useful in improving the resource utilization over pure TDMA systems. In R-MAC systems, subscriber stations (SSs) that wish to transmit into the media have to make a reservation first during the reservation period of the frame. With this in mind, the data-transmission period acts as a pool of data slots (DSs) that can serve any SS reservation. Since the reservation request is smaller in size than the packet size, a better utilization of frame resources is achievable. Another aspect of the suitability of R-MAC for broadband systems is its support for both delay-sensitive (voice) and delay-insensitive (data) information.

A paradox pertinent to the R-MAC protocols is to tune the ratio of the contention and service period resources to serve a better BE traffic performance. In this chapter, we aim at enhancing the R-MAC performance opportunistically through dynamic control of the reservation period. Ultimately, a static allocation ratio that is not adaptive to the variations in traffic load or rate was found to have poorer performance.

There have been a number of proposals for dynamic resource allocation in R-MAC protocols [5–7]. In this work, we institute a framework for an efficient dynamic allocation paradigm to opportunistically maintain and enhance the R-MAC system performance. We describe the desirable characteristics of a dynamically optimized controller that is responsive to the changes in the introduced traffic. In an attempt to implement a resource allocation controller that meets the framework characteristics, we introduce a novel exploratory analysis of the application of dynamic optimization using Markov decision processes (MDP) to leverage the performance of R-MAC protocols. Moreover, we use our dynamic optimization technique with slotted Aloha and p-persistence R-MAC protocols and illustrate considerable performance enhancement. Among existing proposals and other possible ideas, our proposal is a leading work to employ a dynamic optimization tool in adaptively designing the reservation period of R-MAC protocols. Before delving into the details of this proposal, we will acquaint first with the MAC in the IEEE 802.16 standard.

8.2 Multiple Access Protocol of the IEEE 802.16 Standard: Overview

The MAC protocol of the IEEE 802.16 supports both frequency division duplex (FDD) and time division duplex (TDD) of the uplink and downlink portions of the frame. In FDD, the uplink and downlink signals are transmitted at the same time on two separate frequency bands as shown in Figure 8.2. The DL-MAP message defines the usage of the downlink intervals of the frame. This way the SSs could tune into listening to their parts of the time division multiplexed (TDM) data stream transmitted by the base station on the downlink. The UL-MAP message defines the uplink usage of the uplink intervals of the frame. This way the SSs could schedule their transmission during the current frame in the specified time slots.

The MAC frame, which includes the downlink subframe, uplink subframe, and other frame management resources has a fixed length (time duration). A fixed frame length offers the following advantages: (a) facilitates the use of different types of modulation, (b) allows simultaneous use of full-duplex subscriber stations SSs and half-duplex SSs, and (c) simplifies the bandwidth allocation algorithms.

In TDD, both uplink and downlink subframes are transmitted on the same frequency bands but separated in time as shown in Figure 8.3. The TDD frame resources can adaptively be divided between the downlink and uplink subframes as shown in Figure 8.4. This allows better resource utilization during those times when one direction has a heavier traffic load than the other. The downlink subframe is usually scheduled ahead of the uplink subframe, and the DL-MAP message is consequently scheduled ahead of the UL-MAP message as shown in Figure 8.3.

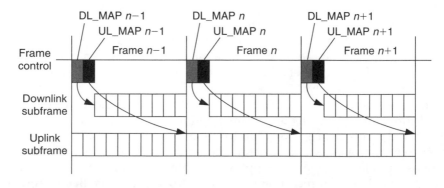

FIGURE 8.2
Frequency division duplex frame organization.

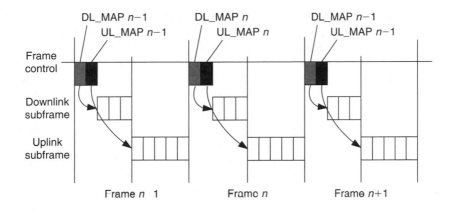

FIGURE 8.3
Time division duplex frame organization.

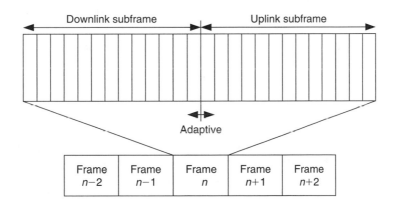

FIGURE 8.4
Adaptive TDD frame.

8.2.1 Downlink Broadcast

The downlink from the base station to the SSs is a point-to-multipoint connection. A central base station, equipped with sectored antenna, broadcasts a TDM into the channel in the direction of the antenna. The base station is the only transmitter operating in this direction during the downlink subframe, so it transmits without having to coordinate with other stations. The receiving stations check the addresses in the DL-MAP message and retain only those messages or data addressed to them.

8.2.2 Uplink Multiple Access

Unlike downlink, the uplink is a multipoint-to-one-point communication link. SSs share the uplink medium in transmitting their data to the base stations. Therefore, organizing the SSs media access is necessary for efficient use of the uplink resources. Access can be broadly categorized into contention-based access and noncontention-based access. The IEEE 802.16 standard adopts a R-MAC in which a combination of demand assigned multiple access (DAMA) and TDMA technologies are employed. R-MAC can support both access categories mentioned above. The uplink subframe is divided into physical slots (PSs). A number of PSs can be grouped into a reservation opportunity (contention slot, CS) or data transmission opportunity (DS) as shown in Figure 8.1. The size of a DS is usually much larger than that of a CS to maintain high throughput. Before delving into the uplink access methods, we first need to understand the different QoS categories and the corresponding scheduling services supported in the uplink subframe of the IEEE 802.16 standard. The standard supports four uniquely characterized scheduling services that correspond to four unique traffic categories [2,4] as follows:

1. *Unsolicited data grants (UGS):* UGS is designed to support real-time service flows that generate fixed-size data packets on a periodic basis, such as voice-over IP traffic. The service offers fixed size unsolicited data grants (transmission opportunities) on a periodic basis. This eliminates the overhead and latency of requiring the SS to send requests for transmission. In UGS, contention-based access is not allowed.

2. *Real-time polling service (rtPS) flows:* rtPS is designed to support real-time service flows that generate variable-size data packets on a periodic basis, such as MPEG video. The service offers periodic unicast request opportunities, which meet the flow's real-time needs and allow the SS to specify the size of the desired grants. The SS is prohibited from using any contention or piggyback requests.

3. *Nonreal-time polling service (nrtPS) flows:* nrtPS is designed to support nonreal-time service flows that require variable-size data grants on a regular basis, such as high bandwidth file transfer protocol (FTP). The service offers unicast request opportunities (polls) on a periodic basis but using more spaced intervals than rtPS. This ensures that the flow receives request opportunities even during network congestion. In addition, the SS is allowed to use contention and piggyback request opportunities.

4. *Best effort (BE) service flows:* The BE service provides fair and efficient service to BE traffic. This is maintained by allowing the SSs to use contention request opportunities. This results in the SS using contention request opportunities as well as unicast request opportunities and unsolicited data grant burst types. SSs access the media through contention. If an SS succeeds in accessing the media, it transmits its reservation request to the base station where the base station responds by assigning the required bandwidth in successive frames on a BE basis. In this chapter, we are primarily

concerned only about the BE traffic and the efficiency of their contention-based access. The rest of this section therefore will shed a spotlight on the contention-based access process.

8.2.3 Reservation Request and Bandwidth Allocation

An SS, which has contention-based BE traffic (data session), must first send a reservation request through contention to the base station. For the rest of the chapter, the term traffic indicates BE traffic. The reservation request or bandwidth request (BWR) contains information about the amount of BW required for the data session transmission. Once a contending SS succeeds in winning contention over one of the CSs, it transmits its BWR to the base station. At this point, it is up to the base station to respond with bandwidth assignment in the subsequent frame(s). The base station (BS) schedules the BE BW allocations according to their order of successful transmissions to the BS.

8.2.4 Contention Resolution Mechanism

The base station organizes the transmission activities on the uplink subframe using the UL-MAP messages. The BS specifies the parts that will be used for contention and data transmissions. Collisions occur in contention intervals when more than one SS wishes to access a CS to transmit a BWR. The standard specifies the use of truncated binary exponential backoff method for contention resolution. When an SS experiences collision, it backs off its retransmission for a number of CSs, which are randomly selected between a minimum and a maximum backoff window values. This random value indicates the number of contention transmission opportunities that the SS shall defer before retransmitting. The minimum and maximum backoff window values are controlled by the base station and distributed to the SSs using a medium access control message transmitted on the downlink.

8.3 Motivation

As mentioned above, this chapter is primarily concerned about the access efficiency of BE traffic. To clearly explain and study the contention and data transmission processes of the BE traffic, we will assume that the entire frame is used solely by BE traffic. Taking this assumption into consideration, Figure 8.5 illustrates the contention and data-transmission processes of the BE traffic. BE traffic uses contention-based reservation to obtain access to the service period of the frame. As shown in Figure 8.5, multiple BWR transmissions in a CS result in a collision. Collided SSs retry transmission of their BWRs in the following CSs. Once a BWR is successfully transmitted to the BS, it enters the service queue. The BWR carries information about the amount of the bandwidth required by the SS. In the IEEE 802.16 standard, the data session

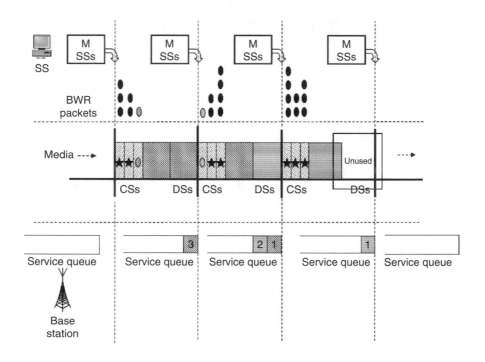

FIGURE 8.5
Contention and data-transmission processes: illustration.

associated with a successful BWR can start transmission in the following frame. For the ease of interpretation in our illustration in Figure 8.5, data transmissions are allowed to start in the same frame in which the BWR was successfully transmitted.

As shown in the second and third frames, the remainder of a data session in a frame resumes transmission in the service period of the next frame. In the third frame, the number of requested DSs in the service queue (one in this example) is less than the number of available DSs (two slots). The result is that a DS will be left unused. Accumulation of such unused DSs would result in poor bandwidth utilization. We argue that letting more BWRs into the service queue increases the throughput of the system. This can be done by increasing the size of the reservation period, which will create more transmission opportunities for the backlogged BWRs. However, since frame time is constant, the reduced service period may result in increased data transmission delay as the service rate will slow down. The contention and data-transmission processes therefore compete for the frame resources. In our study [14], we illustrated the chief importance of the size of the reservation period to the R-MAC performance. On the one hand, increasing the CS resources is highly desirable at times of excessive arrivals of reservation requests (bandwidth request, BWR as we abbreviate throughout this chapter). The consequent decrease in DSs results in an increase in the data-transmission delay and a decrease in system throughput since more resources are used for contention. On the other hand, to remedy the performance, reversing the force by decreasing the contention resources would result in lower number of successful reservations made to the base station resulting in increased contention delay and also decreased throughput. Evidently, network traffic varies over time by characteristic. This motivates us to design a dynamically responsive method to deal with this paradoxical process of contention and service resources allocation under varying traffic load.

8.4   Related Work

R-MAC protocols have been adopted mainly by broadband access networks such as the hybrid fiber cable, DSL, GPRS technologies, and most recently the IEEE 802.16 standard. In this section, an overview of the most prominent reservation multiple access protocols is presented. Then a survey of related studies on performance evaluation and dynamic reservation period allocation policies will be discussed.

8.4.1   Reservation Multiple Access Protocols

A number of R-MAC protocols [16–26] have been proposed. The most popular reservation multiple access protocol is the one based on slotted Aloha [16]. A closely related reservation multiple access protocol is the packet reservation multiple access (PRMA) for local wireless communications [17]. In this protocol, an SS with a data session follows a slotted Aloha contention mechanism to access the media. An SS transmits the first packet of the data session by contention to access the media. Once it successfully accesses a slot, it reserves the same slot in future frames until the end of the data session where that slot is released. In PRMA, data packets that remain for long time in contention are discarded, which is not the case in the reservation Aloha protocol.

Another reservation protocol for mobile communications in a microcellular environment is proposed in [18]. An SS maintains a certain transmission probability to transmit a small reservation packet. If the reservation packet is successfully transmitted, an information packet is assigned to the SS by the base station until the end of the information session. In contrast, if a collision occurs, an adaptive retransmission probability is adopted in a way to maintain system stability. A deviation from PRMA is centralized PRMA (C-PRMA) [19] proposed for a microcellular environment. In C-PRMA, a group of slots is reserved for contention and another group of slots is reserved for data transmission by polling. The base station sends polling information on a slot-by-slot basis. C-PRMA adopts a stack algorithm for reservation to avoid slotted Aloha reservation system instability in higher traffic load environments.

Also the contention-TDMA (C-TDMA) protocol [26], a hybrid of R-Aloha and PRMA, was proposed for radio cellular multi-SS systems. In this protocol, the SSs contend over a slot if it is free. A list of free slots, updated on a frame-by-frame basis, is broadcast by the base station to all the SSs. A CS of a frame in which an SS transmitted the first data packet of its data session, being not in the free slot list in the next frame, indicates that the transmission was successful and that the slot is reserved for the specified SS in consequent frames till the end of the data session. The C-TDMA differs from R-Aloha in that it does not use a broadcast uplink, and it differs from the PRMA protocol in that the slot state is notified by the base station only once per frame, with little overhead.

8.4.2   Performance Evaluation of R-MAC Protocols

The performance of reservation multiple access protocols has been studied in Refs. 5 and 27–35. In Ref. 28, a detailed analysis of reservation-based access system is featured in a GPRS context. The frame under study consists of a contention period at the beginning of the frame followed by a service period, where the ratio between the size of the two periods is statically chosen throughout the time. An SS may transmit a BWR only once during a reservation period, where a CS is chosen at random. Success is determined according to a capture model. An SS whose BWR was not successful in a reservation period waits for the next reservation period and randomly chooses a CS for its BWR transmission. The SSs that successfully transmit their BWRs join a service queue on a first come first served (FCFS) basis. The data packets associated with the successful BWRs are served during the service period, which has a fixed number of DSs. A Markov renewal process embedded at service departure times is utilized in formulating the equilibrium distribution of the number of customers in the system at arbitrary time instances and at customer arrival times.

In Ref. 26, the C-TDMA reservation MAC protocol is studied using a Markovian model. An SS adopts a dynamic permission probability, which is a system parameter, to choose a frame for contention. Once permission is obtained in a frame, a CS is chosen at random. An SS may be in one of the three states: silent, talking, and backlogged. The bidimensional state of the Markov process consists of the number of backlogged SSs and number of transmitting SSs. As the size of the state space of the Markov process does not allow straightforward manipulation, an equilibrium point analysis is used to analyze the Markov process and investigate the delay and throughput performance. In addition, an optimization technique has been proposed to improve the system resources utilization. Upon visiting a certain state, the optimal permission probability is calculated as a function of the number of SSs, number of slots and traffic characteristics. This optimal permission probability is used to calculate the new values of the equilibrium point. This process is recursively iterated until convergence is reached.

The equilibrium point analysis technique was also used in Ref. 18 to evaluate the performance of a proposed R-MAC protocol. The protocol uses a slotted Aloha contention resolution and an adaptive retransmission probability for the purpose of operation stability. The proposed protocol is analytically investigated using a three-dimensional Markov process with the state described by the combination of the number of silent SSs, number of backlogged SSs, and number of SSs, whether waiting in service queue or in transmission. The throughput and delay performance are calculated using the steady-state distribution of the process in equilibrium conditions.

Another performance-evaluation study for an integrated reservation TDMA protocol is conducted in Ref. 29. This study utilizes Markov analysis to compute the contention delay experienced by the BWRs. However, an M/G/1 queuing model is utilized to compute the data-transmission delay spent in the service queue at the base station.

These are the most prominent performance evaluation studies that were cited in the literature in this field. As shown, a common factor among these studies is the use of discrete-time Markov analysis. Although I also utilize Markov analysis in evaluating the system performance, the approach and objective of this chapter are different than these outlined studies. Unlike others, whose approach is mainly to control the transmission probabilities, this chapter emphasizes the design of the reservation period with an objective to control the delay and throughput performance.

8.4.3   Reservation Period Allocation Techniques

Currently, the wide spread of access technologies employing reservation-based multiple access protocols revived the need for optimal protocol operation. The problem of optimal reservation period allocation can be viewed as the key parameter in controlling both delay and throughput performance of the R-MAC protocols. However, there have been a few proposals [7] and Refs. 36–37 reflected on this area. Also Ref. 38 is a comprehensive survey on the contention resolution protocols for the IEEE 802.14 networks. Only recently has the problem of optimal CS allocation gained attention in the literature. There have been a few proposals of dynamic CS allocation algorithms. Sriram and Magil [5] propose a CS allocation algorithm for HFC networks. They observe that a ratio of the number of CSs to that of DSs equal to 3:1 would achieve 100% throughput efficiency in a CS. This observation is based on 33.3% CS throughput efficiency resulting from the use of random binary exponential back-off algorithm for contention resolution. Thus, the rational of the proposed CS allocation algorithm is to keep the reservation period throughput efficiency less than or equal to the available number of DSs in a frame. Otherwise the number of CSs is gradually decreased over the subsequent frames until the ceiling limit of reservation period throughput is restored. The main advantage of this algorithm is to prevent the long-run overflow of BWRs queue at the BS. However, the main concern that can be shared here is the possible excessive increase in contention delay. In the case of relatively high load of BWRs compared to the frame's CS allocation, a scenario that regularly occurs in multiple access networks, high rate of collisions is likely to be experienced. As a consequence, severe contention over the next frames, as a result of the relatively small reservation periods, might lead the contention delay of BWRs beyond acceptable limits of high-speed data communications.

In another proposition, Sala et al. [6] introduce a self-regulating adaptive CS allocation mechanism. All unused DSs are initially allocated as CSs. The self-regulating mechanism dictates that if the number of CSs is too low, the BWRs will not get to the base station, which automatically triggers additional CSs allocation. However, if the number of CSs is too high, more successful requests will reach the base station and the number of empty slots that can be allocated as CSs will decrease. Finally, for any number of CSs allocated in a frame, each SS adopts an optimal transmission probability $p = 1/M_a$, where M_a is the number of active (contending) SSs at the beginning of a CS and $M_a \leq M$, where M is the total number of SSs in the system. This approach requires p to be updated at the beginning of each CS according to the change in the number of active SSs. Adopting smaller transmission probability by SSs has the effect of increasing the probability of a BWR success owing to the decreasing number of transmitted BWRs. Therefore, knowledge of the number of contending SSs is necessary for calculating p. Because the base station BS can attain knowledge only of successfully received BWRs, a pseudo-Bayesian estimator is employed to estimate the average number of active SSs in the next CS using the feedback information resulting from requests transmission in the previous CS. Though this proposal attempts to best utilize the available number of CSs, it has the effect of enlarging the BWR waiting time in the SS queue. From service-delay perspective, it is more advantageous for BWRs to be queued at the base station hoping to gain BW assignment as soon as available resources and service schedules allow.

Another proposal [7] follows a different methodology in dealing with the optimal CSs allocation problem. They aim at maximizing throughput efficiency in the reservation period with an objective to let, through contention, as many BWRs as possible to get into the service queue. To achieve that, they specify that an optimized number of CSs in a frame should equal to the number of BWRs that will be transmitted in that frame. To do that, knowledge of the amount of offered traffic is indispensable. In fact as mentioned earlier, the base station can attain knowledge only about successfully transmitted BWRs. Therefore, two heuristic approaches namely "time proportionality" and "most likely number of requests" are proposed to approximately calculate the average number of initial transmissions and backlogged packet retransmissions, respectively.

As for the performance evaluation studies for the reservation multiple access protocols, the problem of dividing the frame resources between contention and service resources has not yet been comprehensively studied. As we showed in Ref. 14, the size of the reservation period in a frame is a crucial design parameter for the delay and throughput performance of the reservation multiple access protocols.

Regarding the dynamic reservation period allocation proposals presented in the literature, we note that the techniques used to adjust the reservation period size are essentially driven by throughput performance. So, for certain traffic rates and patterns, the delay performance may deteriorate as a result of operating around an optimal throughput point. In telecommunication networks, under limited resources, the delay performance of communications systems retracts after a certain point by increasing the system throughput. Therefore, the delay performance needs to be considered along with throughput performance in allocating the reservation period resources. Such an approach is advantageous for service providers, letting them operate their networks according to various objective metrics. In essence, CSs provide SSs with access opportunities to send their BWRs. Thus, in allocating contention resources, it is crucial to ensure that the level of contention delay does not drive the overall delay (including the data-transmission delay) beyond acceptable figures. In this chapter, we attempt to exploit opportunistic delay and throughput performance gains through dynamic resource allocation administered by the base station, at the beginning of each frame.

8.5 Reservation Period Allocation Controller: Framework

The performance metrics of an R-MAC protocol are the average packet delay and system throughput. Packet delay is defined as the delay spent from the time of initiating a reservation request until the complete transmission of the associated data session. System throughput is defined as the percentage of the system time used for data packets transmission. The objective of the reservation period controller is to optimize the delay or throughput of the R-MAC protocol. The key to optimizing the performance of R-MAC protocol is the control of the ratio between the reservation period and the data-transmission period in the uplink subframe. The inherently varying traffic characteristics of most network environments entail a continuous adjustment of this ratio from one frame to another. Since control information is exchanged between the BS and SSs at the beginnings of the frames, the reservation period controller should also be triggered to work at these time instances. In other words, at the beginning of a frame, the controller is triggered to decide on the ratio between the contention and service intervals that optimize the throughput and delay in that frame. In any given frame, two parameters should be taken into consideration in optimizing this ratio—the number of BWRs in contention and the number of data packets in the service queue of the BS. Based on this information, as well as system performance information (delay and throughput), a base station controller should determine the optimum ratio with the objective of optimizing the long-term delay and throughput performance of the system.

This process governs the dynamic control function of the reservation period. To formally describe the domains of this key function, we formally propose a framework of a reservation period controller as shown in Figure 8.6

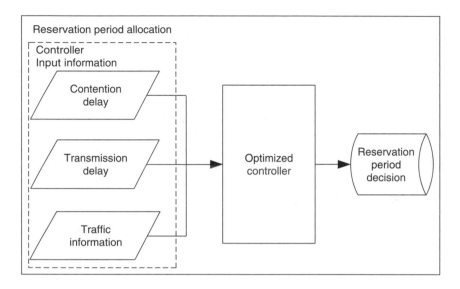

FIGURE 8.6
Block diagram of the proposed framework of the reservation period controller.

and describe its building blocks. In the rest of this chapter, we propose a set of techniques that fulfill the specifications of this framework. However, future research could offer other techniques to provide similar or better results. In what follows, we describe the ideal input information and characteristics of an optimized resource allocation controller for the R-MAC protocol.

8.5.1 Input Information

This information is necessary for the operation of the optimized controller. Contention delay is the average time a BWR has been in contention. Transmission delay is the average time a data packet waits from the time its associated BWR gets successfully transmitted to the BS to the complete transmission of the data packet. The values of these indicators describe how underallocated the reservation period and the transmission period, respectively, are. The third component of the input information is traffic information or in other words traffic-load characteristics.

8.5.2 Optimized Controller Design

The purpose of the optimized controller, as previously stated, is to calculate the optimum reservation period in each frame in an objective to improve the long-term system performance. The reservation period controller should be engaged to operate at the time instances in the beginning of each frame. Any prospective gain in performance would be counterbalanced if the controller's operation was burdensome. Therefore, the optimized controller must possess a set of characteristics to enhance rather than saddle the performance. We state the desirable characteristics of the optimized controller as follows: (a) the decision time is infinitesimally small, (b) the decision is adaptive to performance and traffic conditions, and (c) the controller is tunable to achieve delay and throughput performance improvement.

8.6 Implementation of the Reservation Period Allocation Controller

The methodical realization of the framework components can have many facets. In this section, we present/suggest one way of implementing each of the framework functions as follows.

8.6.1 Input Information Realization

In this section, we investigate the enabling mechanisms to collect contention delay, data-transmission delay, and traffic information.

Contention delay information: In the R-MAC protocols employed in DSL, HFC, GPRS, and IEEE 802.16 standard, the base station has no means to evaluate the contention delay suffered by the BWRs that are contending for the media. We use the number of BWRs that are in contention, denoted B, as a measure of the contention delay. Refs. 6 and 7 apply estimation techniques to compute an average value of B since the BS is incapable of directly recognizing the contending SSs.

Data transmission delay: Accurate information about the number of delays experienced by data packets waiting in the BS's service queue is available at the BS. Similar to the contention delay information, we use the number of waiting data packets, denoted W, as a measure of the data transmission delay.

Traffic information: We characterize the BWR traffic by the probability that an SS transmits a BWR in a CS, and we denote it by P_a. Since the base station is incapable of realizing the traffic characteristics of BWR arrivals, average estimation techniques like pseudo-Bayesian estimator [11] can be employed to estimate P_a at the beginning of the frame.

8.6.2 Optimized Controller

In our proposal, based on an optimization method, the optimized controller processes the input information and reaches a decision (i.e., size of the reservation period), at the beginning of the frame. The chosen size of the reservation period should optimize the utilization of the time slot resources in the frame. We investigate the merits of MDP as the

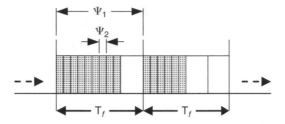

FIGURE 8.7
R-MAC frame with the Ψ_1 and Ψ_2 time units.

core component of the optimized controller in achieving opportunistic performance improvements. MDP is a useful dynamic optimization tool that is based on Markov processes and is often used in control theory, and more generally in decision theory. Ref. 8 is a compact reference on MDP optimization models. In Section 8.7, we describe the MDP optimization model starting by setting up the Markov process of the R-MAC system, whose frame is shown in Figure 8.7. Using the state transition probabilities, we will formulate the MDP model. By solving the optimization problem, we reach the optimal decisions (set of reservation period sizes).

8.7 MDP Optimization Model

As outlined in Section 8.5.1, the input information necessary for the operation of the optimized controller at the beginning of the frame includes the contention delay, data-transmission delay, and traffic information. The number of backlogged BWR packets and number of waiting packets in the service queue at the BS are good representation of the contention- and data-transmission delays, respectively. For traffic characterization, we consider discrete time values of the continuous time transmission probability P_a, as shown in Table 8.1. Therefore, we follow discrete time computations. We introduce the discrete time MDP dynamic optimization method. In MDP, assume a Markov process with a countable state space S. After observing the state of the process, a decision must be chosen. Assume A, which is finite, to be the set of all possible decisions. At the epochs of the state transitions, if the process is in state $i \in S$ and decision $a \in A$ is chosen, then the following occur: (1) a reward $R(i, a)$ is gained, (2) the next state will be chosen according to the transition probabilities $P_{ij}(a)$, where $j \in S$. Thus, both the rewards and the transition probabilities are functions only of the current state and the subsequently chosen decision.

Before examining the suitability of the MDP method for our R-MAC system, we first examine the Markovian property of the R-MAC system shown in Figure 8.7. We let B be the number of BWR packets, which are in contention at the beginning of the frame, and W be the number of waiting data packets,

TABLE 8.1

Discrete Time Values of Continuous Time Ranges of P_a and P_r

Continuous Time Ranges of P_a and P_r	Discrete Time Values of P_a and P_r
$0 \leq P_a < 0.15$	$P_a = 0$
$0 \leq P_a < 0.15$	$P_a = 0.1$
$0.15 \leq P_a < 0.25$	$P_a = 0.2$
$0.25 \leq P_u < 0.35$	$P_a = 0.3$
$0.35 \leq P_a < 0.45$	$P_a = 0.4$
$0.45 \leq P_a < 0.55$	$P_a = 0.5$
$0.55 \leq P_a < 0.65$	$P_a = 0.6$
$0.65 \leq P_a < 0.75$	$P_a = 0.7$
$0.75 \leq P_a < 0.85$	$P_a = 0.8$
$0.85 \leq P_a < 0.95$	$P_a = 0.9$
$0.95 \leq P_a < 1$	$P_a = 1$

at the service queue, at the beginning of the frame. We define the process Ψ_1 with the state space given by $\{[B(f), W(f)] : f \geq 0\}$, which can be shown to be Markovian. The epochs of the state transitions of Ψ_1 are the starting points of the frames. In this study, we are interested in exploiting the dynamic control of the size of the reservation period τ in obtaining opportunistic performance improvements. Since τ can take on a finite set of values, we can think of these values as the domain of decisions. Moreover, since the rate of successful BWR packets and served data packets depend, among other parameters, on τ, the state transition probabilities of Ψ_1 is a function of τ. Based on these similar characteristics, we utilize the MDP method in implementing our proposed optimized controller.

We use another Markov process denoted Ψ_2 to observe—on a CS basis—the arrivals to and departures from the backlog state, as well as the BWR successes that join the service queue. The state transition probabilities of the process Ψ_2 enable the calculation of the transition probability matrix of the process Ψ_1. In what follows, we will derive the transition probabilities of Ψ_1 after describing the working assumptions.

8.7.1 Model Assumptions

Each of M SSs transmits a new BWR in each CS with probability P_a, and retransmits it in each CS, after a collision, with probability P_r. A user can send only one BWR at a time until successfully transmitted, and therefore $0 \leq B \leq M$. This simplifying assumption, used earlier in Ref. 12, assists in keeping track of finitely counted number of backlogged BWRs. However, a user may transmit or retransmit as many times as possible within the same contention period, unlike the assumption of a single transmission per frame in Ref. 3. The service queue, having a finite length of L (i.e., $0 \leq W \leq L$), at the BS, treats the data packets, associated with successful BWRs, in a FCFS fashion. The overflow of data packets is dropped and retransmitted again through reservation [13]. The BWR packet size is equal to the size of a CS. For the simplicity of the model, a BWR packet corresponds to a fixed-size data packet, which has the same size as a DS. The size of a DS is greater than that of a CS; a design principle in R-MAC protocols that leverages throughput performance.

8.7.2 Frame Markov Chain

We observe the state of the process Ψ_1 at fixed and discrete time points at the beginning of the frame and therefore, we call Ψ_1 as the frame Markov chain. As shown in Figure 8.7, the CSs are contiguously organized in the beginning of the frame followed by the DSs (however, other slot organizations are also possible). Since the frame size, denoted by T_f, is constant (as in the IEEE 802.16 standard), knowledge of the number of CSs denoted by τ in a frame implies the number of DSs denoted ε in the same frame, where $\varepsilon = (T_f - \tau \cdot T_{CS})/T_{DS}$, and T_{CS} is the CS time and T_{DS} the DS time.

As shown in Figure 8.7, the time unit of the two-dimensional Markov process Ψ_1, which has a state space given by $\{(B, W); 0 \leq B \leq M, 0 \leq W \leq L\}$, is the frame time T_f. The one-step transition probability matrix of Ψ_1, denoted $^{\Psi_1}P$, has an element value given by

$$^{\Psi_1}P = P\{(B = b_{f+1}, W = w_{f+1})|(B = b_f, W = w_f)\}$$

Computing $^{\Psi_1}P$ is not altogether a straightforward counting problem, and an approach that calculates binomial probabilities of successes and collisions using b_f, τ, P_a, and P_r becomes complicated quickly. Therefore, an enabling mechanism is required to incorporate all combinations of events that may occur during the CS period. In Section 8.7.3, we establish another Markov chain to help us obtain $^{\Psi_1}P$.

8.7.3 Contention Period Markov Chain

We let Ψ_2 denote a two-dimensional Markov process with state space $\{(B, S); 0 \leq B \leq M, 0 \leq S \leq \tau\}$, where B is the number of backlogged users (BWRs) at the beginning of a CS and S the cumulative number of successes since the beginning of the reservation period—be those successes from backlogged or newly transmitted BWRs. As shown in Figure 8.7, the time unit of the process Ψ_2 is the CS. The transition probabilities of Ψ_2, denoted $^{\Psi_2}P$, has an element value given by

$$^{\Psi_2}P = P\{(B = b_f + i, S = s + j)|(B = b_f, S = s)\} \quad \forall 0 \leq i \leq M, 0 \leq j \leq \tau$$

In a similar way to the slotted Aloha analysis in Ref. 11, define $Q_a(i, B)$ as the probability that i out of $M - B$ idle users simultaneously transmit, each with probability P_a, a new BWR in a CS. Also, define $Q_r(i, B)$ as the probability that i out of B backlogged users simultaneously retransmit, each with probability P_r, their backlogged BWRs in a CS.

We have

$$Q_a(i, B) = \binom{M - B}{i} * p_a^i * (1 - p_a)^{M - B - i} \tag{8.1}$$

$$Q_r(i, B) = \binom{B}{i} * p_r^i * (1 - p_r)^{B - i} \tag{8.2}$$

The transition probabilities of the Markov process Ψ_2 can be given by

$$
\Psi_2 P =
\begin{cases}
Q_a(i, B) & 2 \le i \le M - B & j = 0 \\
Q_a(1, B) \cdot [1 - Q_r(0, B)] & i = 1 & j = 0 \\
Q_a(0, B) \cdot [1 - Q_r(1, B)] & i = 0 & j = 0 \\
0 & i = -1 & j = 0 \\
0 & 1 \le i \le M - B & j = 1 \\
Q_a(1, B) \cdot Q_r(0, B) & i = 0 & j = 1 \\
Q_a(0, B) \cdot Q_r(1, B) & i = -1 & j = 1 \\
1 & S = \tau & j = 0
\end{cases}
\tag{8.3}
$$

The last line in Equation 8.3 pertains to invalid transitions out of the absorbing states $(B, \tau) \; \forall \; B$, since only a finite number of CSs, τ, is available in a frame. Now we can obtain the contention results at the end of the reservation period through the τ-step transition probability matrix $\Psi_2 P(\tau)$ using the Markovian property

$$\Psi_2 P(\tau) = \left[\Psi_2 P \right] \tau \tag{8.4}$$

$\Psi_2 P(\tau)$ is useful in tracking all possible combinations of the state evolution over the reservation period, which is in effect an alternative approach to combinatorial techniques in dealing with this problem. Let us now get back to calculating $\Psi_1 P$ using $\Psi_2 P$. The state transitions in Ψ_1 from $(B = b_f, W = w_f)$ to $(B = b_{f+1}, W = w_{f+1})$ entail the following relation:

$$
w_{f+1} =
\begin{cases}
w_f + s - \varepsilon & \text{if } w_f + s > \varepsilon \\
0 & \text{if } w_f + s \le \varepsilon
\end{cases}
\tag{8.5}
$$

From Equation 8.5, knowing w_f, w_{f+1}, and ε, we can determine the S value(s) that cause the indicated state transition. Hence, given the entries of the starting and ending states (b_f, w_f) and (b_{f+1}, w_{f+1}), the entries of $\Psi_1 P$ can be directly obtained from those of $\Psi_2 P$, where we are interested only in elements of $\Psi_2 P$ in which the starting state has $s = 0$. Since the system backlog state is identical in both of the Markov processes Ψ_1 and Ψ_2, the entries of $\Psi_1 P$ can be given by Equation 8.6. We will frequently use the terms $\Psi_2 P_{u,v}^{(x)}$ and

$^{\Psi_z}P_{u,v}$ to denote the x-step and one-step transition probabilities, respectively, from state u to state v of the Markov process Ψ_z.

$$
^{\Psi_1}P = \begin{cases}
\displaystyle\sum_{y=0}^{\tau} {}^{\Psi_2}P^{(\tau)}_{(b_f,0)(b_{f+1},y)} & w_{f+1} = 0 \quad \varepsilon \geq L \\[2em]
\displaystyle\sum_{y=0}^{\min(\tau,\varepsilon-w_f)} {}^{\Psi_2}P^{(\tau)}_{(b_f,0)(b_{f+1},y)} & w_{f+1} = 0 \quad \varepsilon < L \\[2em]
0 & w_{f+1} > 0 \quad w_f + s > L \\[1em]
\displaystyle\sum_{y=s}^{\tau} {}^{\Psi_2}P^{(\tau)}_{(b_f,0)(b_{f+1},y)} & w_{f+1} > 0 \quad w_f + s = L \\[2em]
{}^{\Psi_2}P^{(\tau)}_{(b_f,0)(b_{f+1},s)} & w_{f+1} > 0 \quad w_f + s < L \\[1em]
0 & \text{otherwise}
\end{cases}
\tag{8.6}
$$

where the first line is the case when the total number of data packets at the service queue, resulting from the sum of w_f data packets already waiting from the previous frame and s successful BWRs associated with which an equal number of data packets, will all be transmitted in frame f. The second line is similar to the first one but with discarding the possibilities when S drives the total number of waiting data packets beyond the number of DSs ε in the frame. The third line is for those invalid transitions that require a total number of waiting data packets at the end of the reservation period that exceeds the size of the service queue L. The fourth line, for the case when the service queue is full at the end of the reservation period, considers the possibility of additional successes that were dropped out of the service queue owing to its finite length. The fifth line is straightforward. Thus, similar to the MDP model, it is obvious from the entries of Equation 8.6 that the state transition probabilities are functions only of the present state (b_f, w_f) and the subsequent decision τ. In Ref. 14, we use the frame Markov model in evaluating the delay and throughput performance of a similar R-MAC system. After calculating the transition probabilities of the Markov process Ψ_1 that describes the frame behavior, we are now ready to formulate the MDP optimization model.

8.7.4 Optimization Problem Formulation

A policy determines the rule used to choose a decision, that is, in our study the decision is the size of the reservation period, should the system be in a specific state [8]. In broadband networks it should be up to the service provider's operational objectives to set that policy. That will depend on the local environment in terms of the traffic characteristics. Since the traffic pattern varies from one environment to the other, a reservation period allocation policy has not been endorsed by any of the broadband network standards employing R-MAC protocols.

If we let X_f denote the state of the Markov process at time f (at the beginning of frame f) and θ_f be the decision (size of reservation period) chosen at time f, then the transition probability from state $u = (b_f, w_f)$ to state $v = (b_{f+1}, w_{f+1})$ is given by

$$P\{X_{f+1} = v | X_0, \tau_0, X_1, \tau_1, \ldots\ldots, X_f = u, \tau = \theta_f\} = P_{u,v}(\theta_f)$$

For any policy β, the limiting probability $\pi_u^{\theta_f}$ that the process will be in state u and decision θ_f will be chosen is given by

$$\pi_u^{\theta_f} = \lim_{f \to \infty} P_\beta\{X_f = u, \tau = \theta_f\} \tag{8.7}$$

where $\pi_u^{\theta_f}$ must satisfy the following:

(a) $\pi_u^{\theta_f} \geq 0 \quad \forall u, \theta_f$

(b) $\sum_u \sum_{\theta_f} \pi_u^{\theta_f} = 1$

(c) $\sum_{\theta_f} \pi_v^{\theta_f} = \sum_u \sum_{\theta_f} \pi_u^{\theta_f} P_{u,v} \quad \forall v$

where $P_{u,v}$ is taken from the entries of $\Psi_1 P$. When the system is in state $u = (b_f, w_f)$ and decision θ_f is taken, a reward $R(u, \theta_f)$ is achieved. Using $\pi_u^{\theta_f}$, we can calculate the steady-state expected reward as follows:

$$E[R] = \lim_{f \to \infty} E[R(u, \tau_f) | u, \theta_f] = \sum_u \sum_{\theta_f} \pi_u^{\theta_f} R(u, \theta_f) \tag{8.8}$$

Consequently, the optimal policy that maximizes the expected average reward with respect to θ_f is

$$\underset{\pi = \pi_u^{\theta_f}}{\text{maximize}} \sum_u \sum_{\theta_f} \pi_u^{\theta_f} R(u, \theta_f) \tag{8.9}$$

subject to the following conditions:

1. $\sum_{\theta_f} \pi_u^{\theta_f} = \sum_v \sum_{\theta_f} \pi_v^{\theta_f} P_{v,u}(\theta_f)$

2. $\sum_u \sum_{\theta_f} \pi_u^{\theta_f} = 1$

3. $\pi_u^{\theta_f} \geq 0$

where the policy $P_u^{\theta_f}$ (the probability of taking decision θ_f when the process is in state u) is computed from the following equalities:

$$\pi_u^{\theta_f} = \pi_u P_u^{\theta_f} \quad \text{and} \quad \sum_{\theta_f} \pi_u^{\theta_f} = \pi_u$$

The maximum average reward can be achieved by a nonrandomized policy [8], that is, the decision that must be taken when the process is in state u is a deterministic function of u. We now need to design a reward function such that controlled-state transitions, through dynamic tuning of the size of the reservation period would yield performance enhancement.

8.7.5   Reward Function

The reward function $R(u, \theta_f)$ is a function of the current state $u = (b_f, w_f)$, and subsequently the chosen size of the reservation period θ_f. The reward function should be designed such that desirable state transitions result in higher rewards than undesirable transitions. After observing the state at the beginning of a frame to be $u = (b_f, w_f)$, the reward function shall be calculated for all the values in the domain of the decision, which corresponds to the size of the reservation period. Finally, the chosen decision is the one that achieves the highest return of the reward function.

According to the framework in Section 8.5, the choice of a reservation period should consider improving the performance opportunistically. To comply with this requirement, and since throughput performance is inseparable from the delay performance, we use a multiobjective reward function structure. We let $R_D(u, \theta_f)$ denote the delay objective function and $R_{th}(u, \theta_f)$ denote the throughout objective function. Ultimately, $R(u, \theta_f)$ is a parametric function of $R_D(u, \theta_f)$ and $R_{th}(u, \theta_f)$. For this multiobjective optimization, we choose the aggregation function technique [9] to aggregate all the objectives into a single function using a form of weighted sum as follows:

$$R(u, \theta_f) = g_D \cdot R_D(u, \theta_f) + g_{th} \cdot R_{th}(u, \theta_f) \tag{8.10}$$

where $0 \le g_D \le 1$ and $0 \le g_{th} \le 1$ are the weighting coefficients representing the relative influence of the delay and throughput components.

8.7.5.1   Delay Objective Function

Two parts comprise the delay performance: contention delay and data-transmission delay. As we have mentioned, improving one part usually results in deterioration of the other part. Therefore, the two components must be separated and represented in the reward function. Thus, Equation 8.10 is slightly expanded as follows:

$$R(u, \theta_f) = g_c \cdot R_{D_c}(u, \theta_f) + g_w \cdot R_{D_w}(u, \theta_f) + g_{th} \cdot R_{th}(u, \theta_f) \tag{8.11}$$

where $R_{D_c}(u, \theta_f)$ and $R_{D_w}(u, \theta_f)$ pertain to contention and data-transmission delays, respectively, with $0 \le g_c \le 1$ and $0 \le g_w \le 1$ as their respective weight coefficients.

8.7.5.1.1 Contention Delay Objective Function

Our objective is to reduce the number of BWR packets that are in contention at the beginning of the next frame. As the size of the reservation period increases, more BWR packets are more likely to leave the contention state. Meanwhile, the reward function should consider the relative value of B in deciding on a good value of θ_f. Accordingly, we propose the following experimental form of contention delay objective function.

$$R_{D_c}(u, \theta_f) = 1 - \exp\left(\frac{-\theta_f}{b_f}\right) \tag{8.12}$$

The proposed function in Equation 8.12 has good characteristics of an objective reward function. For numerical illustrations, we take $M = 15$ and $L = 30$, and $\tau = \{3, 6, 9, 12, 15, 18, 21, 24, 27\}$ is the domain of the size of the reservation period. As shown in Figure 8.8, the proposed objective function has desirable reward differentiation according to the combination of b_f and θ_f.

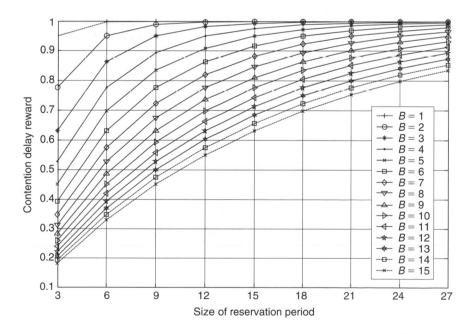

FIGURE 8.8
Contention delay reward differentiation with different combinations of B and τ.

8.7.5.1.2   Data-Transmission Delay Objective Function

Similar to the contention delay reward function, the data-transmission delay objective reward function is

$$R_{D_w}(u, \theta_f) = 1 - \exp\left(\frac{-\theta_f}{w_f}\right) \tag{8.13}$$

8.7.5.2   Throughput Objective Function

Given the state of the process $u = (b_f, w_f)$ at the beginning of a frame, we need to calculate the throughput of that frame for $\tau = \theta_f$. We let $E\lfloor N_{v_f} | b_f, w_f \rfloor$ denote the expected number of data packets that were served in frame f. The frame throughput is defined as the effective fraction of the frame time utilized in data packets transmission. Therefore, the average conditional throughput of frame f, th_f, is given by

$$E[th_f | b_f, w_f] = \frac{T_{DS}}{T_f} E[N_{v_f} | b_f, w_f] \tag{8.14}$$

Typically, N_{v_f} is expressed as

$$N_{v_f} = \begin{cases} w_f + s_f & w_f + s < \varepsilon \\ \varepsilon & w_f + s \geq \varepsilon \end{cases} \tag{8.15}$$

In Equation 8.15, based on our early assumption of fixed and equal size of data packets, the maximum number of served data packets in a frame cannot exceed the number of DSs in that frame. Otherwise, the number of served data packets is the sum of waiting data packets at the beginning of the frame and the number of data packets that join the service queue in the same frame, that is, the number of successful BWR packets. Accordingly, the conditional expected number of served data packets in a frame is computed as

$$E[N_{v_f} | B = b_f, W = w_f] = \begin{cases} w_f + \sum\limits_{b_{f+1}=0}^{M} \sum\limits_{s=0}^{\varepsilon - w_f - 1} s\ \Psi_2 P_{(b_f, 0)(b_{f+1}, s)}^{(\tau)} & w_f < \varepsilon \\ \varepsilon & w_f \geq \varepsilon \end{cases} \tag{8.16}$$

where the first and second terms in Equation 8.16 correspond to the first and second cases of Equation 8.15, respectively. By direct substitution in Equation 8.14, the conditional frame throughput is

$$E[th_f | B = b_f, W = w_f] = R_{th}(u, \theta_f) = \begin{cases} \dfrac{T_{DS}}{T_f}\left[w_f + \sum\limits_{b_{f+1}=0}^{M} \sum\limits_{s=0}^{\varepsilon - w_f - 1} s\ \Psi_2 P_{(b_f, 0)(b_{f+1}, s)}^{(\tau)} \right] & w < \varepsilon \\ \dfrac{T_{DS}}{T_f}\varepsilon & w \geq \varepsilon \end{cases}$$

$$\tag{8.17}$$

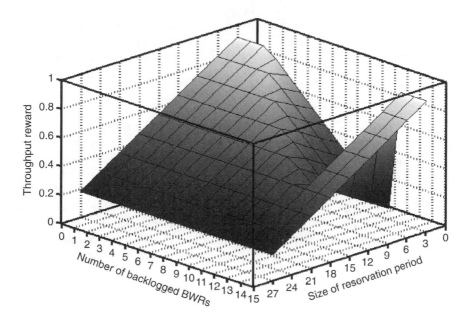

FIGURE 8.9
Throughput reward differentiation for different combinations of B and τ with $W = 12$.

The frame throughput function in Equation 8.17 has appropriate character-istics of an objective reward function because (a) its maximum value is 1, (b) it is a function of the current state $u = (b_f, w_f)$ and chosen decision θ_f, and (c) it responds appropriately to the different state transitions that result in different throughput rewards. Hence, we directly use the formula in Equation 8.17 as an objective function. As shown in Figure 8.9, we plot the throughput objective function for $W = 12$ and $P_a = 0.3$ to show sample reward differentiations. In Figure 8.9, we note that as B increases, the reward is highest at a median value of τ. This is desirable to maintain efficient contention and data-transmission processes. We also note that as B decreases, the reward tends to be highest at lower values of τ, which is also desirable to efficiently serve the waiting data packets in the service queue. Plugging Equations 8.12, 8.13, and 8.18 in Equation 8.11, we obtain

$$
R(u, \theta_f) = g_c \cdot \left[1 - \exp\left(\frac{-\theta_f}{b_f} \right) \right] + g_w \left[1 - \exp\left(\frac{-\theta_f}{w_f} \right) \right]
$$

$$
+ g_{th} \cdot \begin{cases} \dfrac{T_{DS}}{T_f} \left[w_f + \displaystyle\sum_{b_{f+1}=0}^{M} \sum_{s=0}^{\varepsilon - w_f - 1} s \ \Psi_2 P_{(b_f,0)(b_{f+1},s)}^{(\tau)} \right] & w < \varepsilon \\[4ex] \dfrac{T_{DS}}{T_f} \varepsilon & w \geq \varepsilon \end{cases} \tag{8.18}
$$

We are now ready to solve the optimization problem for the optimal decision that should be chosen, should the process be in any state. We use a relative value iteration algorithm [10] to solve the MDP with average reward.

8.7.6 Implementation Complexity

MDP are known for their computational complexity with large-scale state space systems [15]. Therefore, real-time calculation of the optimum size of the reservation period at the beginning of the frame would be infeasible. Instead, the optimization problem shall be solved offline for each discrete time value of the transmission probability shown in Table 8.1. The result will be the optimum size of reservation period conditioned on the system state at the beginning of the frame (b_f, w_f) and the estimated transmission probability P_a. The resulting optimized decisions (optimum values of θ_f) will be entered into a lookup table, which is administered by the base station (a sample lookup table is shown in Figure 8.10).

8.7.7 Operation of the Optimized Controller

In operation, after observing the state of the system (b_f, w_f) and estimating the average transmission probability P_a, the controller retrieves, from the precalculated lookup table, the corresponding size of the reservation

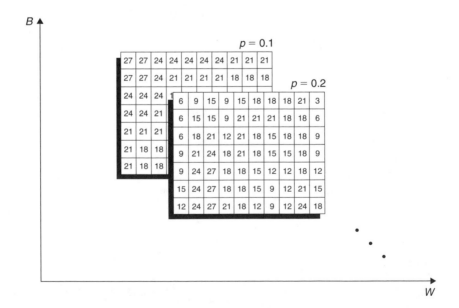

FIGURE 8.10
Multiple lookup tables.

period of the starting frame. This method results in considerable performance enhancements as we shall see in the numerical examples in Section 8.8.

8.8 Performance Evaluation

We conduct an illustrative analysis for a relatively small size R-MAC system. In our simulation, the base station deterministically collects the values of B, W, P_a, and P_r at the beginning of each frame. Using the lookup table, the controller determines the value of τ to be used in that frame. We consider $M = 15$, $L = 30$, and T_f to be 1 ms resembling the IEEE 802.16 recommendation, however, we consider a downscaled number of slots per frame, 33 slots/frame. We let the size of a reservation request packet be one slot, and we let the size of a data packet be three slots (although we consider a fixed-size data packet, our model can be slightly modified to accommodate variable-size data packets). Thus, a frame can have a maximum of 33 CSs or alternatively a maximum of 10 DSs. The size of the reservation period will have the domain $\tau = \{3, 6, 9, 12, 15, 18, 21, 24, 27\}$. We implement our optimization model in two contention resolution environments; the slotted Aloha and p-persistence mechanisms and obtain performance results as follows.

8.8.1 Slotted Aloha Contention Resolution

The objective here is to study the performance enhancements resulting from our optimization model under constant and variable values of P_a. We implement the BWR traffic intensity by varying the value of P_a. However, P_r remains unchanged considering it to be a system parameter. In this example, we show the transient R-MAC performance degradation as a result of the increase in P_a. We let the simulation run enough time to reach the steady state. Taking $P_r = 0.1$, we let $P_a = 0.15$ up till the 10th second. We find that under static reservation period allocation, $\tau = 15$ achieves the best throughput performance among all values of static allocations. In Figure 8.11, up till the 10th second, the best throughput performance under static allocation approaches 44.7%, which is the same as the throughput bound derived in Ref. 6. During the same period, our optimized controller achieves 49% throughput, which represents about 10% throughput improvement. Meanwhile, the throughput enhancement results in increased packet delay as shown in Figure 8.12. At the 10th second, we increase P_a to 0.3 until the end of the simulation time. As a result, the system throughput under both static and dynamically optimized allocations is slightly affected; however, our optimized allocation maintains the same 10% throughput improvement. The performance is only slightly affected by doubly increasing P_a, first, because the unchanged retransmission probability partially absorbs the increase in P_a and second, because by assumption we do not let B grow indefinitely by restricting the generation of new BWRs to those SSs in idle state.

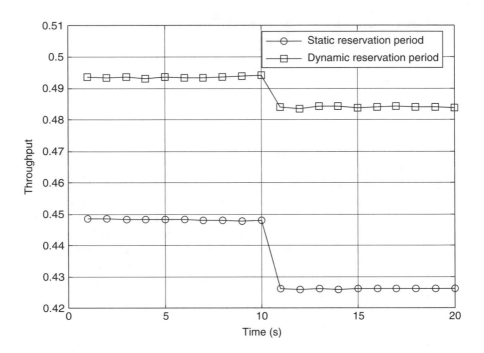

FIGURE 8.11
Slotted Aloha throughput with increasing P_a at the 10th second.

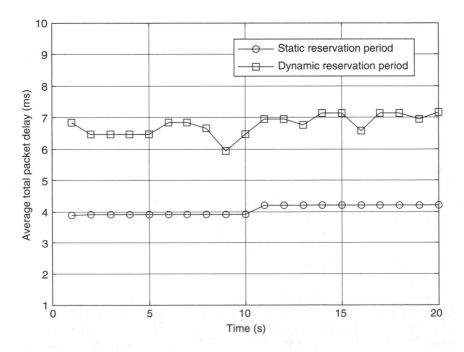

FIGURE 8.12
Slotted Aloha packet delay with P_a increasing at the 10th second.

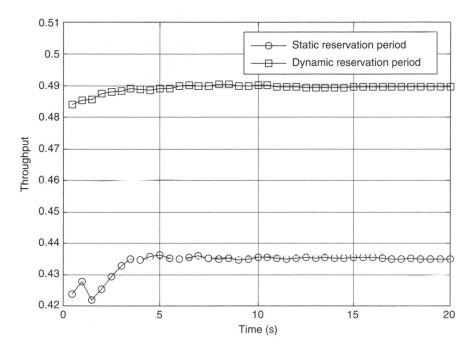

FIGURE 8.13
Slotted Aloha throughput with P_a frequent oscillation.

In Figure 8.13, instead of alternating P_a only once, we continuously choose between $P_a = 0.15$ and $P_a = 0.3$ by tossing a coin with equal probability of returning heads or tails. Once one value is chosen, it remains active for a geometrically distributed number of frames, with an average of 50 frames. As shown in Figure 8.13, the optimized controller consistently outperforms the throughput of the best case static allocation by about 12%. The corresponding delay increase is shown in Figure 8.14. Figures 8.15 and 8.16 show the dynamically optimized allocation of the reservation period, in an interval of 200 frames around the 10th second of the simulation time, under the two traffic scenarios. In Figure 8.15, in response to the double increase of P_a from 0.15 to 0.3 at the 10th second, the optimized controller adapts quickly by increasing the reservation period to $\tau = 21$ more frequently. In Figure 8.16, the more frequent alternation of P_a between 0.15 and 0.3 the deeper the dynamicity of the adaptive response and the more visits to the decision $\tau = 21$.

8.8.2 *p*-Persistence Contention Resolution

Using traffic scenarios similar to the slotted Aloha case, we investigate the performance improvement achievable through our optimized controller. In the

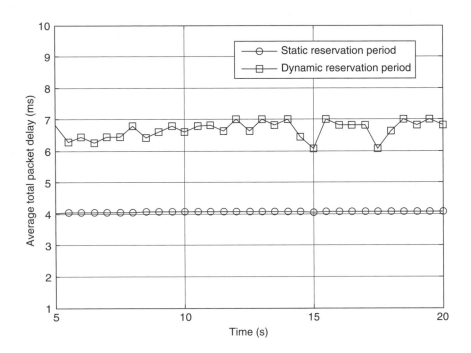

FIGURE 8.14
Slotted Aloha packet delay with P_a frequent oscillation.

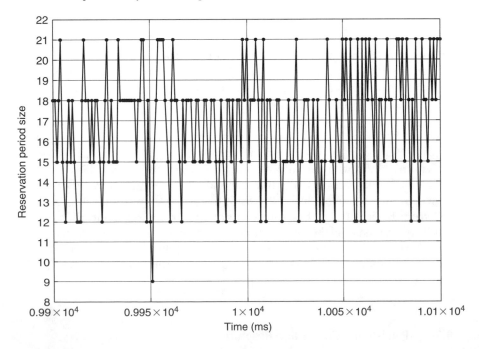

FIGURE 8.15
Optimized controller transient response around the 10th second.

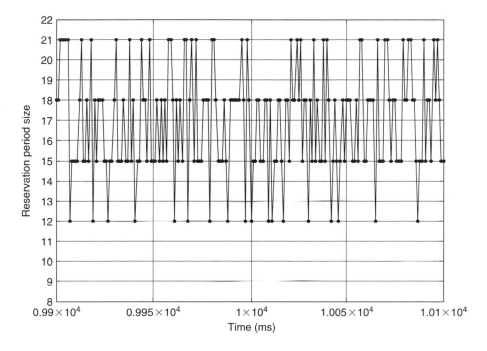

FIGURE 8.16
Optimized controller response to P_a frequent oscillation.

p-persistence mechanism, each SS persistently attempts to transmit a BWR with the same probability in every CS. Hence, P_a and P_r are the same and are denoted as p. The first traffic scenario is designed to observe the transient behavior associated with the increase in p from 0.15 to 0.3 at the 10th second of the simulation time. As shown in Figure 8.17, under both optimized dynamic and best case static reservation period allocation, the increase in the BWR traffic results in sudden deterioration in both allocation policies. However, the dynamically optimized performance is better by about 30% than the best case static allocation $\tau = 15$, with the delay increase shown in Figure 8.18. Moreover, when we alternate between $p = 0.15$ and $p = 0.3$ throughout the simulation time in the same way as described in Section 8.8.1, the dynamically optimized performance is about 44% better than the best case static allocation $\tau = 15$ from a throughput perspective as shown in Figure 8.20. The corresponding delay increase is shown in Figure 8.21. In Figure 8.19, the transient response of the optimized controller to the steep performance degradation is shown. The controller adapts to the traffic increase by consistently maintaining a wide reservation period. Figure 8.19 shows the controller's activity in an interval of 200 frames around the 10th second of the simulation time. Also Figure 8.22 shows the change activity of the controller decisions, in an interval of 200 frames around the 10th second of the simulation time, where

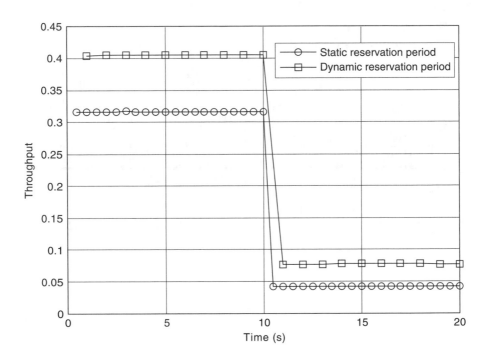

FIGURE 8.17
p-Persistence throughput with increasing p at the 10th second.

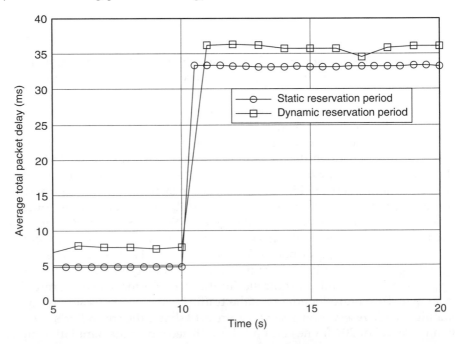

FIGURE 8.18
p-Persistence packet delay with increasing p at the 10th second.

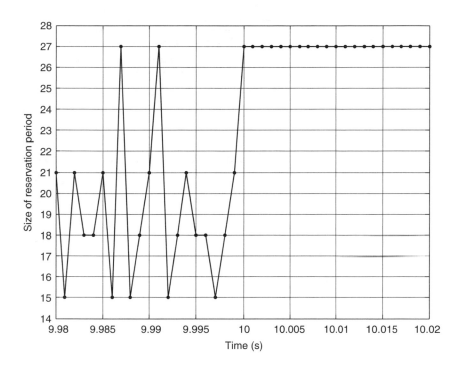

FIGURE 8.19
Optimized controller transient response around the 10th second.

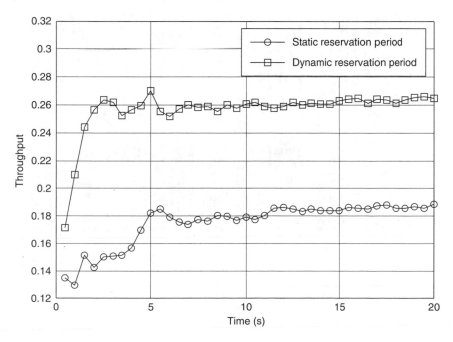

FIGURE 8.20
p-Persistence throughput with *p* frequent oscillation.

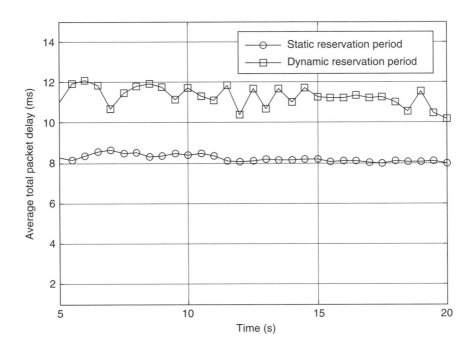

FIGURE 8.21

p-Persistence packet delay with *p* frequent oscillation.

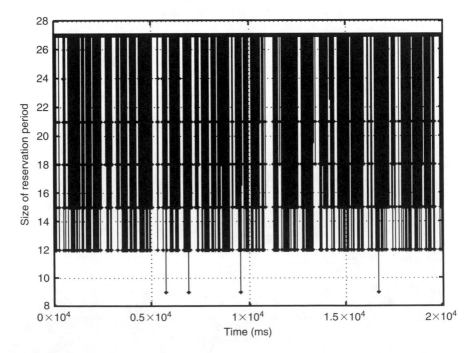

FIGURE 8.22

Optimized controller response to *p* frequent oscillation.

the widest reservation period $\tau = 27$, in the set of possible decisions, is taken more often.

8.9 Conclusions

The telecommunications market is constantly reshaping. The recent ratification of the IEEE 802.16 standard (WiMAX) will enable the telecom players to offer bundled fixed and mobile broadband services more efficiently. The growing interest in the IEEE 802.16 revitalized the interest to enhance the R-MAC performance. The dynamic characteristics of today's traffic may degrade the R-MAC performance. A paramount design parameter in R-MAC systems is the size of the reservation period. Through proper control, tuning the size of the reservation period according to the traffic and operations dynamics leads to considerable performance enhancement. We established a framework of an ideal reservation period controller, administered at the base station, which can be embraced by different implementation methods. Building on the Markovian behavior of the R-MAC system, we used two-stage Markov processes to formulate a MDP model, which resembles the dynamics of the R-MAC protocol. Using our MDP model, we established a method to dynamically calculate the optimized size of the reservation period at the beginning of each frame given the traffic information and state of the system. To illustrate the merits of the proposed method, we examined our optimization model under the slotted Aloha and p-persistence contention resolution mechanisms. With both techniques we obtained considerable throughput enhancement over the static reservation period policy. Our proposed optimization method uses an offline-calculated lookup table, which alleviates the MDP computational complexities, and hence adds no delay to the R-MAC performance. Furthermore, our method meets the proposed framework's operational characteristics. Since it does not require any change in the standards, it can be directly employed in the MAC control of the DSL, HFC cable, GPRS, and the IEEE 802.16 technologies in compliance with their specifications. The results we have obtained in this chapter indicate a potential enhancement in the R-MAC performance over previously derived performance bounds in Ref. 6. Therefore, for future research, there is a need to investigate bounds on the throughput performance obtainable through our dynamic optimization model.

References

1. IEEE 802.16 standard, IEEE standard for local and metropolitan area networks—part 16: air interface for fixed broadband wireless access systems, 2004.
2. C. Eklund, R. B. Marks, K. L. Stanwood, and S. Wang, IEEE 802.16 standard: A technical overview of the wireless MAN air interface for broadband wireless access, *IEEE Communications Magazine*, June 2002.

3. J. N. Daigle and M. N. Magalhaes, Analysis of packet networks having contention-based reservation with application to GPRS, *IEEE/ACM Transactions on Networking*, vol. 11, Issue: 4, 2003, pp. 602–615.

4. A. Ghosh, D. R. Wolter, J. G. Andrews, and R. Chen, Broadband wireless access with WiMax/8O2.16: current performance benchmarks and future potential, *IEEE Communications Magazine*, 2005, pp. 129–136.

5. K. Sriram and P. D. Magil, *Enhanced Throughput Efficiency by Use of Dynamically Variable Request Minislots in MAC Protocols for HFC and Wireless Access Networks*, Kluwer Academic Publishers, Telecommunication Systems Journal, vol. 9, Issue: 3, 1998, pp. 315–333.

6. D. Sala, J. Limb, and S. Khaunte, Adaptive control mechanism for cable modems MAC protocols, *Proceedings Infocom Conference* 1998, San Francisco, PA, March 1998.

7. W. M. Yin and Y. D. Lin, Statistically optimized minislot allocation for initial and collision resolution in hybrid fiber coaxial networks, *IEEE Journal on Selected Areas in Communications*, vol. 18, Issue: 9, Sept. 2000, pp. 1764–1773.

8. S. M. Ross, *Applied Probability Models with Optimization Applications*, Dover Publications, Mineola, NY, 1970.

9. C. Coello, A. Carlos, V. Veldhuizen, A. D. Lamont, and B. Gary, *Evolutionary Algorithms for Solving Multi-Objective Problems*, Kluwer Academic Publishers, Boston, MA, 2002.

10. Institut National de la Recherche Agronomique, http://www.inra.fr/bia/T/MDPtoolbox/index_category.html.

11. D. Bertsekas and R. Gallager, *Data Networks*, 2nd ed. Prentice Hall, Englewood Cliffs, NJ, 1992.

12. R. Rom and M. Sidi, *Multiple Access Protocols: Performance and Analysis*, Springer Verlag, Berlin, 1990.

13. A. Doha and H. Hassanein, Access delay analysis in reservation multiple access protocols for broadband local and cellular networks, *29th Annual IEEE International Conference on Local Computer Networks*, 2004, pp. 752–759.

14. A. Doha, H. Hassanein, and G. Takahara, Performance evaluation of reservation medium access control in IEEE 802.16 networks, *IEEE International Conference on Computer Systems and Applications*, 2006, pp. 369–374.

15. C. H. Papadimitriou, and J. N. Tsitsiklis, The complexity of Markov decision processes, *Mathematics of Operations Research*, vol. 12, Issue: 9, 1978, pp. 441–450.

16. W. Crowther, R. Rettberg, D. Walden, S. Ornstein, and F. Heart, A system for broadcast communication: reservation-Aloha, *Proceedings of the 6th Hawaii International Conference on System Sciences*, 1973, pp. 596–603.

17. D. J. Goodman, R. A. Valenzuela, K. T. Gayliard, and B. Ramamurthi, Packet reservation multiple access for local wireless communications, *IEEE Transactions on Communications*, vol. 37, Issue: 8, 1989, pp. 885–890.

18. N. M. Mitrou, T. D. Orinos, and E. N. Protonotarios, A reservation multiple access protocol for microcellular mobile-communication systems, *IEEE Transactions on Vehicular Technology*, vol. 39, Issue: 4, 1990, pp. 340–351.

19. G. Bianchi, F. Borgonovo, L. Fratta, L. Musumeci, and M. Zorzi, C-PRMA: a centralized packet reservation multiple access for local wireless communications, *IEEE Transactions on Vehicular Technology*, vol. 46, Issue: 2, 1997, pp. 422–436.

20. D. J. Goodman, Cellular packet communications, *IEEE Transactions on Vehicular Technology*, vol. 38, Issue: 8, 1990, pp. 1272–1280.

21. D. Tsai and J.-F Chang, A hybrid contention based TDMA technique for data transmission, *IEEE Transactions on Communications*, vol. 36, Issue: 2, 1988, pp. 225–228.

22. C. G. Kang, C. W. Ahn, K. H. Jang, and W. S. Kang, Contention-free distributed dynamic reservation MAC protocol with deterministic scheduling (C-FD3R MAC) for wireless ATM networks, *IEEE Journal on Selected Areas in Communications*, vol. 18, Issue: 9, 2000, pp. 1623–1635.

23. A. Sugihiara, K. Enomoto, and I. Sasai, Hybrid contention/reservation channel-access schemes for integrated voice/data wireless networks, *Seventh IEEE International Symposium on Personal, Indoor and Mobile Radio Communications PIMRC'96*, vol. 2, Oct. 1996, pp. 638–642.

24. J.-F. Frigon, V. C. M. Leung, and H. C. B. Chan, Dynamic reservation TDMA protocol for wireless ATM networks, *IEEE Journal on Selected Areas in Communications*, vol. 19, Issue: 2, 2001, pp. 370–383.

25. H. C. B. Chan, J. Zhang, and H. Chen, A dynamic reservation protocol for LEO mobile satellite systems, *IEEE Journal on Selected Areas in Communication*, vol. 22, Issue: 3, 2004, pp. 559–573.

26. G. Pierobon, A. Zanella, and A. Salloum, Contention-TDMA protocol: performance analysis, *IEEE Transactions on Vehicular Technology*, vol. 51, Issue: 4, 2002, pp. 781–788.

27. S. Tasaka, Stability and performance of the R-Aloha packet broadcast system, *IEEE Transactions on Computers*, vol. C-32, 1983, pp. 717–726.

28. J. N. Daigle and M. N. Magalhaes, Analysis of packet networks having contention-based reservation with application to GPRS, *IEEE/ACM Transactions on Networking*, vol. 11, Issue: 4, 2003, pp. 602–615.

29. Y. Cao, H. R. Sun, and K. S. Trivedi, Performance analysis of reservation media-access protocol with access and serving queues under bursty traffic in GPRS/EGPRS, *IEEE Transactions on Vehicular Technology*, vol. 52, Issue: 6, 2003, pp. 1627–1641.

30. R. Fantacci and S. Nannicini, Performance evaluation of a reservation TDMA protocol for voice/data transmission in microcellular systems, *IEEE Journal on Selected Areas in Communications*, vol. 18, Issue: 11, 2000, pp. 2404–2416.

31. K. Crisler and M. Needham, Throughput analysis of reservation Aloha multiple access, *Electronics Letters*, vol. 31, Issue: 2, 1995, pp. 87–89.

32. D. G. Jeong and W. S. Jeon, Performance of an exponential backoff scheme for slotted Aloha protocol in local wireless environment, *IEEE Transactions on Vehicular Technology*, vol. 44, Issue: 3, 1995, pp. 470–479.

33. T. K. Liu, J. A. Sivester, and A. Ploydoros, Performance evaluation of R-Aloha in distributed packets radio networks with hard real-time communications, *IEEE 45th Vehicular Technology Conference*, vol. 2, July 1995, pp. 554–558.

34. F. L. Presti and V. Grassi, *Markov Analysis of the PRMA Protocol for Local Wireless Networks*, J.C. Baltzer AG, Science Publishers, Amsterdam.

35. C. S. Wu and G.-K. Ma, Performance of packet reservation MAC protocols for wireless networks, *IEEE 48th Vehicular Technology Conference*, vol. 3, 1998, pp. 2537–2541.

36. G.-H. Hwang and D.-H. Cho, Dynamic random channel allocation scheme in HiperLAN Type 2, *IEEE International Conference on Communications*, 2002, ICC 2002, vol. 4, pp. 2253–2257.

37. Z. J. Haas and J. Deng, On optimizing the backoff interval for random access schemes, *IEEE Transactions on Communications*, vol. 51, Issue: 12, pp. 2081–2090.
38. N. Golmie, Y. Saintillan, and D. Su, A review of contention resolution algorithms for IEEE 802.14 networks, *IEEE Communications Surveys*, vol. 2, Issue: 1, 1999, pp. 2–12.

9

Scheduling Algorithms for OFDMA-Based WiMAX Systems with QoS Constraints

Raj Iyengar, Koushik Kar, Biplab Sikdar, and Xiang Luo

CONTENTS

9.1 Introduction

The IEEE 802.16 is an emerging suite of standards for point-to-multipoint (PMP) broadband wireless access (BWA). The 802.16e amendment to the 802.16-2004 specification enables support for combined fixed and mobile operation for licensed and license-exempt frequencies below 11 GHz. IEEE 802.16 is likely to emerge as a preeminent technology for cost-competitive ubiquitous BWA supporting fixed, nomadic, portable, and fully mobile operations offering integrated voice, video, and data services. The technology is likely to be considered in a variety of deployment scenarios, such as standalone IP core-based networks and as a data overlay over existing broadband and cellular networks. Initial deployments are likely to be based on fixed/nomadic operation with fully mobile usage to follow. Of the three different physical layers (PHYs) specified in the standard, orthogonal frequency division multiple access (OFDMA) is likely to emerge as the most preferred PHY supporting all usage models, owing to the inherent flexibility it offers to the system designer.

Consequently, we focus on scheduling/resource allocation problems over an OFDMA PHY. In this chapter, we focus on single 802.16–based cells, although in practice, it is likely that multiple cells will coexist, resulting in more complicated and interesting challenges.

The frame structure typically used in IEEE 802.16–based wireless systems is shown in Figure 9.1. The initial portion of the frame (control) consists of the downlink map (DL-map) and the uplink map (UL-map). These specify information about the allocations made for each client on uplink/downlink. Broadly speaking, these maps contain information about which subchannels and time slots are allocated to a given user in a given frame. The downlink portion of the frame is followed by the uplink portion. The horizontal axis denotes time and the vertical denotes subchannels used in OFDMA (hence this axis denotes frequency). Figure 9.1 shows a time division duplex (TDD) frame, for an OFDMA PHY, with allocations made for three users on the

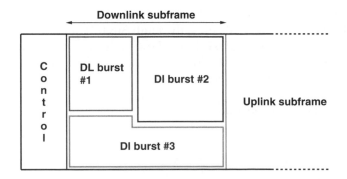

FIGURE 9.1
Typical frame structure in wireless scenarios.

downlink subframe. For the remainder of the chapter, we assume that the system operates over an OFDMA PHY layer.

We note that the IEEE 802.16 standard specifies a number of techniques for users to request bandwidth resources on the uplink, and a suite of QoS options. We focus on the specific QoS requirement that the user requires an allocation, which allows the user to transfer a certain amount of data over a certain time interval. This is applicable to the unsolicited grant service (UGS) and real-time polling service (rtPS) classes of QoS specifically.

The scheduling problem is to allocate time slots on a subset of the subchannels available (frequency resource) to meet client's demands and maximize system throughput. allocations are not classified on the basis of burst profiles, but allocations are made on a user basis. The time interval T, over which these demands must be satisfied, can be equal to the frame duration T_F or some other value, and can be interpreted as a time horizon over which the QoS requirements must be met. For example, if the channel conditions are quickly changing, then T can be assumed to be a small value, of the order of a few frame times. However, if the channel conditions are slowly varying, T can be a larger value. Practically, however, it is more likely that the former case will be encountered.

9.1.1 Contributions

There is very little literature on scheduling algorithms for wireless networks built around the IEEE 802.16 [2] standard. The draft [2] specifies a number of hooks and features that can be used, but does not specify the exact scheduling algorithm to allow vendors differentiate their products. Note that in a centralized wireless network, the scheduler is one of the most important components of the system. To the best of the authors' knowledge, no published work using the techniques and algorithms described in this chapter exist. While a number of papers on the problem of *bit loading* for OFDMA systems exists, these works do not explore in detail the combinatorial nature of the problem (e.g., refer to Ref. 5). In contrast, in this chapter, we provide a proof of hardness for the discrete version of the resource allocation problem (described later) and a *provably good* algorithm for the LP relaxation based on solutions to mixed covering and packing LPs. We also present a heuristic based on generalized concurrent flow, which performs well in numerical experiments. We discuss the extensions for formulations that take into account power control, as well as mixed integer program formulations for overhead efficient allocations. Finally, we also provide formulations and solution techniques for scenarios under more constricting assumptions on the user radios, and multihomed clients in a network scenario.

9.1.2 Organization of This Chapter

In Section 9.2, we describe the system model that is used in the LP formulations. The LP formulations are presented in Section 9.3. Section 9.3.1 discusses the solution to a simple case of the problem, when all channel conditions are identical. In Section 9.3.3, we prove that the discrete version of the resource allocation problem is NP-hard. Sections 9.3.4 and 9.3.5 discuss two different approaches to solve the continuous time relaxation of the discrete version of the problem. Section 9.3.5 presents a heuristic approach, which is analyzed numerically in Section 9.3.6. In Section 9.4, we discuss a simplified version of the power-control problem and show how we can extend the earlier formulations to solve the power and slot allocation problem. We note that the formulations have nonlinear constraints and are likely very difficult to solve. Section 9.5 concludes the chapter and presents some interesting directions for future work, which this work has thrown open.

9.2 System Model

The time and frequency resources that must be shared between clients are represented in Figure 9.2. This serves as an abstraction of the OFDMA PHY used. In the remainder of this section, we will describe how the system model described in the initial portion of the section relates back to specific 802.16 features. Alternatively, we describe what features of 802.16 can be used to provide the information required as input to the scheduling algorithm. Unlike an OFDM system, the OFDMA system provides the added flexibility of allocating a subset of available carriers to a user for some time duration. In Figure 9.2, the time axis (horizontal) is discretized into slots of length Δ. The notion of a *slot* is discussed in more detail in Section 9.2.2. The vertical axis represents the different subchannels used in the system. We note that a *subchannel* is a logical entity, which is comprised of elemental subcarriers. This is discussed in more detail in Section 9.2.1. The channel conditions perceived by

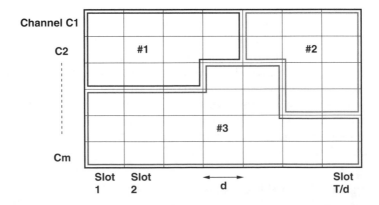

FIGURE 9.2
Representation of time and frequency resources.

each station are captured in a channel conditions matrix of dimension $n \times m$ (m subcarriers and n users). The entries in this matrix are a measure of the rate achievable by user i on subcarrier j. For example, these entries may have units of bits/second, which is intuitively useful since the allocations for each user are time durations on channels.

9.2.1 Frequency Diverse and Frequency Selective Scheduling

IEEE 802.16 allows for different mappings of subcarriers, which are the elemental frequency-level components of an OFDM/OFDMA system to subchannels, which are logical groupings of subcarriers. An example of a frequency selective mapping is partially utilized subcarrier (PUSC), where the subcarriers that constitute a subchannel are randomly selected from all over the available bandwidth. Owing to the inherent diversity of the constituents of a subchannel, the conditions seen by any user on any given subchannel are roughly the same. In the case of *Band AMC*, the subcarriers that constitute a subchannel are adjacent to each other, and the channel conditions seen by a user vary across all the subchannels and further, these vectors vary across users also. Note that, we do not consider the impact of schemes used for the purpose of interference averaging. In the remainder of the chapter, we focus on scheduling/resource allocation problem applicable to systems over a PUSC and band AMC-based OFDMA for 802.16. For the purposes of this chapter, we do not distinguish between fully utilized subcarrier (FUSC) and PUSC. We use the abstraction that the distribution of the constituent subcarriers of each subchannel results in some interesting properties that can be exploited during the scheduling/resource allocation process. An introduction to subchannelization techniques can be found in Ref. 1. In the next section, we consider specific versions of the scheduling problem that are applicable to the two broad techniques of subcarrier to subchannel mapping.

9.2.2 Notion of Slot at Physical Layer

Physical slot (PS) is the basic unit of time at the PHY layer. In the case of OFDMA, this corresponds to a symbol time. A number of PSs together constitute a *slot*. Since resource allocation is typically a MAC layer problem, we focus on the slot and subchannel allocation problem. We note that the techniques developed in this chapter are applicable even when measurements are made at the individual subcarrier level, although in practice this is unlikely, partly because of the overhead that the system will incur.

9.2.3 Channel Quality Indication

The 802.16 draft specifies that certain channels on the uplink are designated as channel quality indication channels (CQICH). Clients provide average CINR (carrier to interference-plus-noise ratio) measures feedback that they perceive on the downlink using this channel. The base station specifies a CQICH allocation for a particular client, in the control portion of the frame, which instructs the client to provide the average CINR measure feedback using the fast feedback channel to the base station. The measurement of channel quality on the downlink is in itself an interesting issue when there are a large number of clients, since channel quality feedback overhead becomes an important factor in the scheduling efficiency. We assume for the purpose of this work that channel quality information is collected for all clients at the same time, and at regular intervals, using either the CQICH or some similar mechanism. Uplink channel conditions can be estimated at the base station every time data are sent out from a client. In the case of AMC operation, we assume that information feedback is in the form of a vector of channel measurements for each user across all subchannels in the system.

9.2.4   UGS and rtPS QoS Classes

The algorithms presented in this chapter are applicable to the UGS and rtPS QoS classes defined in the 802.16 draft standard [2]. We consider the following implementation of the UGS class: each user requires the capability to transfer a fixed minimum amount of data over a given time interval T. We note that there is an inherent assumption that it is sufficient to find or be given as input a value T over which the possible satisfaction of demands is acceptable to the applications being used at the user, or as defined in a form of a service-level agreement (SLA) with the user. We use an identical notion of the rtPS service, with implicit assumptions on the periodicity of the polling intervals for the minimum rate required by an rtPS flow.

9.3   Problem Formulation: Frequency and Time Allocation with QoS Constraints

In this section, we present LP formulations for the version of the scheduling (resource allocation) problem in the absence of the power allocation considerations. The objective of the formulation is to maximize the sum throughput across all users so that the demand for each user is satisfied. The motivation for this formulation is that it captures the (in general) conflicting objectives of the network (maximizing throughput, as seen in the objective function) and the users (satisfying demand).

We consider two versions of the resource allocation problem, one where the time axis is continuous and another where the time axis is discretized as shown in Figure 9.2. We show later that the discrete version of the problem is in general NP-hard, which motivates the study of the continuous time relaxation.

In the discrete case, the resource allocation problem reduces the allocation of each of these slots to individual stations to achieve certain objectives. In the continuous relaxation, the resource allocation problem is to allocate time chunks to users across the available subcarriers to satisfy demand and maximize throughput.

In Equations 9.1 through 9.4, users perceive different conditions on each subcarrier and these values differ across users. Clients modulate multiple subcarriers concurrently. The set of subcarriers allocated to a user is typically a subset of the total number of subcarriers available in the system. Let α_{ij} represent the rate achievable by user i on channel j in bits/second. For the current problem formulation, we assume that the α_{ij} are simple functions of the CINR vectors fed back to the base station by the users. We note that this imposes restrictions on the power allocated by a user on a subchannel. For the case of this problem formulation, it can be assumed that the user splits traffic in a static manner (e.g., equally) across the subcarriers allocated to it.

$$\max \sum_{i=1}^{n} \sum_{j=1}^{m} \alpha_{ij} x_{ij} \tag{9.1}$$

$$\text{subject to: } \sum_{i=1}^{n} x_{ij} \leq T \quad \forall j = 1, \ldots, m \tag{9.2}$$

$$\sum_{j=1}^{m} \alpha_{ij} x_{ij} \geq d_i \quad \forall i = 1, \ldots, n \tag{9.3}$$

$$x_{ij} \geq 0 \tag{9.4}$$

The formulation in Equations 9.1 through 9.4 is referred to as LP(1) in this document. In the above formulation, the variables x_{ij} represent the time duration allotted to station i on channel j to transmit data. The exact position of this time chunk is communicated to the user by the base station using control messages that are broadcast to all users at the start of each frame (these are referred to as downlink map and uplink map in the standard [2], respectively). The objective function seeks to maximize the overall amount of data in bits transmitted. Without any demand constraints, this problem can be solved simply as discussed a little later. The first constraint specifies that the total time allocated across all stations on a channel cannot exceed the duration T. The second constraint is the QoS constraint and specifies that the total data transmitted by a station i in time T must at least equal the demand d_i in bits, in case of uplink traffic. In the case of downlink traffic, this is the minimum amount of data that must be received by station i. Note that during time duration, T is a time horizon over which

QoS guarantees must be provided. Note that in these situations there is an inherent assumption that channel conditions do not vary significantly over the update interval, when channel condition updates are sent from the client to the base station, in the case of downlink traffic. In the case of uplink traffic, uplink channel conditions can be measured at the base station roughly every T seconds.

The LP(1) is the relaxation of the integer program shown below:

$$\max \sum_{i=1}^{n} \sum_{j=1}^{m} \alpha_{ij} n_{ij} \Delta \tag{9.5}$$

$$\text{subject to: } \sum_{i=1}^{n} n_{ij} \leq \frac{T}{\Delta} \quad \forall j = 1, \ldots, m \tag{9.6}$$

$$\sum_{j=1}^{m} \alpha_{ij} n_{ij} \geq \frac{d_i}{\Delta} \quad \forall i = 1, \ldots, n \tag{9.7}$$

$$n_{ij} \geq 0 \tag{9.8}$$

$$n_{ij} \in Z^{+} \tag{9.9}$$

Here, n_{ij} are the number of slots allocated to station i on channel j. It is assumed that the slot length Δ exactly divides the subframe (uplink or downlink) time T (as shown in Equation 9.2).

9.3.1   Identical Channel Conditions

We first address the simple case where all users perceive identical channel conditions. Consider the discrete version of the resource allocation problem where cells need to be assigned to users so that demand is satisfied and throughput is maximized. A simple algorithm can be used to achieve these objectives. The simplifying factor is the fact that all users perceive identical channel conditions; hence, allocating a slot on one channel is the same as allocating a slot on any other channel to a user. Hence, we can simply allocate one slot at a time in a round-robin fashion among the users, till demand is satisfied. Note that when the demand of all users is satisfied, the algorithm terminates with some slots left unallocated. These can be allocated arbitrarily. It is easy to see that in the case where the demands of the stations are not satisfiable, the algorithm returns an allocation that is max–min fair. We note that this is applicable for the case of PUSC/FUSC scenarios, with roughly identical channel conditions for each user on any given subchannel. We further note that practically, subcarrier distribution across the band is not guaranteed to ensure equal channel conditions as assumed. It is an interesting open question to investigate the actual performance hit that the system will take as a consequence of the identical channel assumption.

9.3.2 Choice of T

As mentioned earlier, T is the time horizon over which QoS guarantees must be provided. The formulations shown so far require each user to be allowed to transmit a certain number of *bits* every T time. This is an example of a rate constraint that must be satisfied for a user. For example, the users could be mapped to a guaranteed bandwidth MAC-level QoS service, which can be used to provide T1 services to homes and small businesses. Typically, it can be assumed that T is at most a few frames long. Alternatively, a different optimization problem can be formulated to minimize the total time horizon over which user demands can be satisfied. This formulation will look like:

$$\min T \tag{9.10}$$

$$\text{subject to:} \quad \sum_j \alpha_{ij} x_{ij} \geq d_i \quad \forall i \tag{9.11}$$

$$\sum_i x_{ij} \leq T \quad \forall j \tag{9.12}$$

$$x_{ij} \geq 0 \tag{9.13}$$

We note that the form of this linear program allows for a solution using techniques described later in this chapter.

9.3.3 Hardness Result

In this section, we prove that the demand-constrained discrete version of the problem is NP-hard. In the discrete version of the problem, slots on each channel must be assigned to clients (a slot can be assigned only to a single client) so that demands are satisfied and throughput is maximized. The proof is by reduction from MAXIMUM CONSTRAINED PARTITION (henceforth referred to only as PARTITION), which is a well-known NP-complete problem, for example, refer to Ref. 7 or 8. For completeness, we state the problem here. An instance of PARTITION is a finite set A and a size $s(a) \in Z^+$ for each $a \in A$. A solution to an instance of PARTITION is a partition of A, a subset $A' \subseteq A$, so that

$$\sum_{a \in A'} s(a) = \sum_{a \in A - A'} s(a) \tag{9.14}$$

(The optimization version of this problem seeks to maximize the number of elements from S on the same side as a given element a_0.) Now consider the following version of the discrete scheduling problem. There are some number of subcarriers, m, and each subcarrier has only one time slot associated with it. There are only two clients, both of which see exactly the same channel conditions on the given set of channels (assuming that the channel conditions seen by the two clients can be represented as integers). Hence, we have a set A consisting of m elements, each having some value a_i, $i = 1, \ldots, m$. Let each user has a demand $d = \sum_i a_i/2$. Since each element of the set can be assigned to only one client and not more, we see that we can solve this problem iff we can solve the PARTITION problem. Therefore, even this simplified version of the discrete scheduling problem is NP-complete. Hence, we can say that the general discrete scheduling problem described earlier (for throughput maximization) is NP-hard.

9.3.4 An Input-Dependent Approximation Algorithm for LP(1)

In this section, we present an input-dependent approximation algorithm for the LP in Equations 9.1 through 9.4 based on results for approximating mixed covering and packing linear programs. In Ref. 6, the author describes efficient sequential algorithms to solve the feasibility problem approximately. Specifically, the algorithm returns a solution satisfying all constraints within a $1 \pm \epsilon$ factor in $O(Mp \log M/\epsilon^2)$ time, where M is the number of constraints and p the maximum number of constraints any variable occurs in.

It is possible to use the efficient feasibility algorithms as a subroutine to calculate the optimal solution by using a bisection search on the range of the optimal solution. Assuming we know the maximum data rate achievable across all channels in the system (denoted as W), we can compute an approximately optimal solution in $O(k \log mWT)$ time, where k is the time complexity for a single call to the approximate feasibility subroutine.

In the case of the LP(1), we note that $M = m + n + mn$ (supply, demand, and nonnegativity constraints, respectively), and p can have a value of at most 5 for a given iteration. This can be seen as the demand constraint for a given client $i = 1, \ldots, n$ contains one occurrence of x_{ij} for some $j = 1, \ldots, m$. Similarly, the supply constraint for channel j contains one occurrence of x_{ij}. When performing a bisection search, two constraints are added for a given range of the objective function, each containing one occurrence of the variable x_{ij} (the fifth occurrence is because of the nonnegativity constraint). In reality, since the number of orthogonal carriers (or subcarriers in OFDMA) for a given system is typically fixed, as n grows large, the complexity looks roughly like $O(n \log (n)/\epsilon^2)$.

Note that the upper bound on the system throughput can be improved in the following way: If the subcarriers are numbered $1, \ldots, m$, denote s_i as the station with the best data rate on subcarrier i. Let W_{s_i} be the maximum rate achievable across all users on subcarrier i. Therefore, the running time of the algorithm can be improved to $O(k \log \sum_{i=1}^{m} W_{s_i} T)$.

9.3.5 A Heuristic Approach Based on Maximum Concurrent Flow

In this section, we present a heuristic for the LP in Equations 9.1 through 9.4, which makes use of the maximum concurrent flow interpretation of the LP in Equations 9.1 through 9.4. The advantage of this heuristic is that its time complexity does not depend on the value of the maximum data rate achievable on a given channel. An alternate formulation of the relaxed resource allocation problem (Equations 9.1 through 9.4) can be to maximize a common multiple of satisfied demand across all users, that is, some λ so that at least λd_i is satisfied for all i. However, this is not a traditional concurrent flow problem. There are multipliers associated with some of the variables in the formulation. This can be posed as a *generalized flow* problem. Efficient techniques to approximate generalized concurrent flow are presented in Ref. 9. The path formulation of the generalized concurrent flow [9] problem is:

$$\max \lambda \tag{9.15}$$

$$\text{subject to:} \quad \sum_{P:e\in P} \gamma_P(e)x(P) \leq u(e) \quad \forall e \tag{9.16}$$

$$\sum_{P\in P_j} x(P) - \lambda d_j \geq 0 \quad \forall j \tag{9.17}$$

$$x(P) \geq 0 \quad \forall P \tag{9.18}$$

$$\lambda \geq 0 \tag{9.19}$$

Given an s–t (source–destination) path $P = e_1, \ldots, e_r$, $\gamma_P(e_q) = 1/\Pi_{i=q}^{r}\gamma(e_i)$. This is the amount of flow sent into arc e_q to deliver one unit of flow at t using path P. For the formulation in Equations 9.1 through 9.4 under consideration, we can construct a graph with mn paths, with m paths from one source of data through m channels connected to a sink. The variables y_{ij} can be interpreted as bits. Note that only the edges corresponding to the m channels have a capacity of T associated with them, the rest of the edges are uncapacitated. The resulting formulation is shown in Equations 9.20 through 9.24.

$$\max \lambda \tag{9.20}$$

$$\text{subject to:} \quad \sum_{i=1}^{n} \frac{y_{ij}}{\alpha_{ij}} \leq T \quad \forall j = 1, \ldots, m \tag{9.21}$$

$$\sum_{j=1}^{j=m} y_{ij} - \lambda d_j \geq 0 \quad \forall i = 1, \ldots, n \tag{9.22}$$

$$y_{ij} \geq 0 \quad \forall i, j \tag{9.23}$$

$$\lambda \geq 0 \tag{9.24}$$

The intuition behind this heuristic is represented diagrammatically in Figure 9.3.

Once the concurrent flow problem is solved, the optimal λ will either be larger or smaller than 1. In case of the former, we are dealing with an infeasible program, and the resulting allocations are in fact a good solution to the resource allocation problem. In case of the latter, the program is feasible but the resulting allocation is in general not throughput optimal. In this case, scale the solution back by the objective function value in the generalized concurrent flow program. Next, allocate the remainder of the frame in a throughput optimal manner, by allocating the remaining time on each subcarrier to the client with the best data rate on that subcarrier. From the results in Ref. 4, the time complexity of the concurrent flow heuristic is $O[\epsilon^{-2}(k_1 + k_2 \log k_2)(k_1 + n)]$, where $k_1 = 2mn + m$ is the number of edges and $k_2 = 2(m + n)$ the number of nodes in the graph on which the concurrent flow is computed. We note that

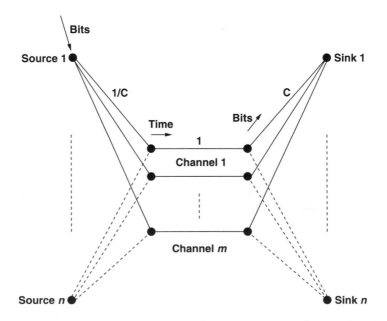

FIGURE 9.3
Concurrent flow formulation.

the advantage of this formulation is that the algorithm is not dependent on the values in the channel conditions matrix. The solution provided by the heuristic is such that some portion of the frame is used for the concurrent flow-based allocation; that is, when the optimal solution is scaled back, the total time allocated on each subcarrier is the same. This is because the optimal solution to the concurrent flow problem will find allocations x_{ij} so that $\sum_i x_{ij} = T \ \forall j$. This is true because any leftover space would imply a larger value of λ.

9.3.6 Numerical Results

In this section, we present simulation results for the heuristic outlined in Section 9.3.5. We assume an 802.16 system with frame time 5 ms. All clients have the same demand in these experiments. The channel conditions for each client are chosen randomly between 1 and 10 Mbp. The system operates over five subcarriers. ϵ is chosen to be small, $\epsilon = 0.01$, so that the concurrent flow problem is almost optimally solved. The optimal solution is computed using CPLEX to solve the linear program. We note that the heuristic performs very well for the problem instances considered, and closely approximates the optimal solution, from Figures 9.4 and 9.5.

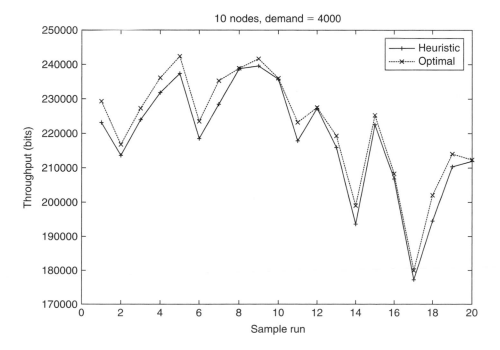

FIGURE 9.4
Performance of heuristic, 10 nodes, demand = 4000.

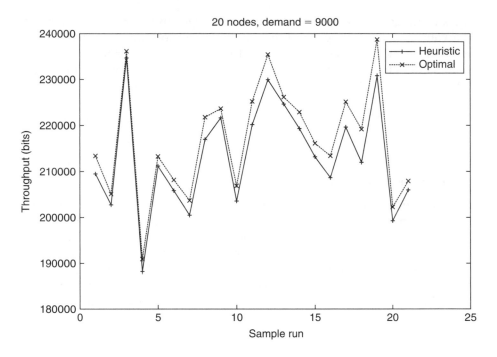

FIGURE 9.5
Performance of heuristic, 20 nodes, demand = 9000.

9.4 Joint Channel and Power Allocation

We note that the solution to the throughput maximization formulation implicitly does not consider the power allocation problem. In general, this is not optimal from a power-control perspective. In this section, we formulate a power-control optimization problem to allocate power for each user optimally across different subcarriers/tones, while trying to satisfy QoS constraints. The objective function in this formulation remains the same, we seek to maximize throughput, subject to demand constraints for users.

$$\max \sum_i \sum_j r_{ij} x_{ij} \tag{9.25}$$

$$\text{subject to:} \quad r_{ij} = B \log \left(1 + \frac{p_{ij}}{n_{ij}} \right) \tag{9.26}$$

$$\sum_j P_{ij} \leq P_{\max} \quad \forall i \tag{9.27}$$

$$\sum_{j} r_{ij} x_{ij} \geq d_i \qquad (9.28)$$

$$\sum_{j} x_{ij} \leq T \qquad (9.29)$$

$$x_{ij}, r_{ij} \geq 0 \qquad (9.30)$$

The new terms introduced in this formulation are described below:

- p_{ij}: power allocated by user i on channel j.
- r_{ij}: rate achieved by user i on channel j.
- n_{ij}: interference and noise power seen by user i on subchannel j.
- R_i: The total rate for a user across all channels must be larger than the target rate R_i for a given user i.
- $P_{\max}$: limit on power for a user.

We note that the above formulation (Equations 9.25 through 9.30) is difficult to solve owing to the presence of constraints of the form $r_{ij} x_{ij}$. Given the difficult nature of the above power-control problem, we focus on a simplified version, which is amenable to analysis in extremal SINR regimes. In this setup, a set of orthogonal channels must be assigned to a set of users, where each user splits its power optimally across the channels allocated to it. While the optimal power allocation solution has a "water-filling" type structure, the optimal channel assignment problem is very challenging owing to the non-linear dependence of user throughput on the set of channels assigned to it. Since the optimal channel allocations are computationally intensive to obtain in general, we analyze the system in the two extremal SINR regimes (very high and very low SINR) and show how the optimal solutions can be obtained in these regimes in a computationally efficient manner. Finally, we demonstrate that the best of the optimal solutions obtained for the two extremes show excellent (close to optimal) performance over the entire SINR range. We note that the nature of the power-control problem considered here is applicable for uplink communication. Without loss of generality, we assume that time is slotted and focus on the channel allocation problem across users for a single given time slot.

Recall that multiple channels can be assigned to a single user, but a single channel cannot be shared by multiple users. Therefore, a valid assignment of subchannels to users corresponds to a one-to-many mapping from users to channels. In this chapter, we refer to such an assignment as a *polymatching* in the user-channel bipartite graph (see Figure 9.6). Let Φ denote the set of all polymatchings in the user-channel graph. The throughput of user i on a channel j assigned to it is of the form $B_j \log(1 + \kappa p_{ij}/n_{ij})$, where B_j and κ are constants. For ease of exposition, we will assume $B_j = B \; \forall j$ and $\kappa = 1$, although the analysis and algorithms that we present in this chapter can easily

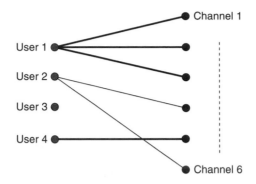

FIGURE 9.6
A polymatching: the figure shows one valid polymatching for four users and six channels. (Note that the polymatching is represented by the bold edges.)

be extended to the more general case. The throughput maximization problem for the entire system can be posed as

$$\max \sum_{i=1}^{L} \sum_{j:(i,j)\in\phi} \log\left(1 + \frac{p_{ij}}{n_{ij}}\right) \tag{9.31}$$

$$\text{subject to: } \sum_{j\in\phi_i} p_{ij} \leq P_i \quad \forall i \tag{9.32}$$

$$p_{ij} \geq 0 \quad \forall i \quad \forall j \tag{9.33}$$

$$\phi \in \Phi \tag{9.34}$$

Note that for a given polymatching ϕ, the above problem reduces to the optimal power allocation problem for each user, whose solution corresponds to a "water filling" across the different channels assigned to the user. The problem posed above is much more complex; however, since it corresponds to a joint channel and power allocation problem, it requires us to find the channel assignment (polymatching) that will yield the best system throughput under optimal power allocations for that channel assignment. A naive approach to solve this problem would be to enumerate all polymatchings, compute the attainable throughput for a polymatching by running the water-filling algorithm, and then pick the polymatching that yields the maximum throughput value. However, since the number of polymatchings is in general exponential in the size of the user-channel bipartite graph, this naive approach is computationally very expensive, and not feasible for large number of channels or users. Thus our goal is to obtain optimal or near-optimal channel assignments in a computationally efficient manner.

9.4.1 Throughput Analysis in the High SINR Regime

In this section, we analyze the throughput attained by a user i in the high SINR regime to motivate an algorithm that computes a channel assignment (polymatching) that is optimal in this regime. Consider any user i, and let $\phi_i = \{j : (i,j) \in \phi\}$ denote the set of channels assigned to user i. Let $k_i = |\phi_i|$. In the high SINR regime, $P_i \gg n_{ij} \ \forall j \in \phi_i$. This implies that if power allocation is done to optimize user throughput, the user will use all the channels assigned to it, and the power allocations will correspond to a water-filling solution, as characterized by

$$p_{ij} + n_{ij} = \lambda_i \quad \forall j \in \phi_i \tag{9.35}$$

Summing over all the k_i channels, we obtain

$$\lambda_i = \frac{P_i + N_i}{k_i} \tag{9.36}$$

where P_i is the aggregate transmission power of user i and N_i, the *aggregate noise power* of user i, is defined as $N_i = \sum_{j \in \phi_i} n_{ij}$.

If we represent the throughput attained by user i as U_i, then we can write

$$U_i = \sum_{j \in \phi_i} \log \left(1 + \frac{p_{ij}}{n_{ij}} \right) \tag{9.37}$$

$$= \sum_{j \in \phi_i} \log \left(\frac{p_{ij} + n_{ij}}{n_{ij}} \right) \tag{9.38}$$

$$= \sum_{j \in \phi_i} \log \left(\frac{P_i + N_i}{k_i n_{ij}} \right) \tag{9.39}$$

$$\approx \sum_{j \in \phi_i} \log \left(\frac{P_i}{k_i n_{ij}} \right) \tag{9.40}$$

where the approximation comes from the fact that in the high SINR regime, $P_i \gg N_i$. Hence, we can write the throughput for user i as

$$U_i = k_i \log(P_i) - k_i \log(k_i) - \sum_{j \in \phi_i} \log(n_{ij}) \tag{9.41}$$

The importance of writing this expression for the throughput is that it allows us to quantify the *incremental utility* of allocating the jth channel to user i. In general, this incremental utility is a function of i, j, k_i.

Specifically, consider the incremental utility of allocating channel j to user i, when $k-1$ channels have already been allocated to it. Then, using the throughput expression in Equation 9.41, the incremental utility in this case, denoted by α_{ijk}, is expressed as follows (note that in both cases—before as well as after the allocation of channel j—the total power P_i available at user i is divided using water-filling technique across the set of allocated channels):

$$\alpha_{ijk} = \log(P_i) - \left[k\log(k) - (k-1)\log(k-1)\right] - \log(n_{ij}) \qquad (9.42)$$

In the above expression, $(k-1)\log(k-1)$ at $k=1$ should be interpreted as 0.

We note that the expression for incremental utility expression includes only terms specific to the added channel j ($\log(n_{ij})$), the number of channels already allocated (k), and the total power of user i, P_i. Thus, the incremental utility expression does not depend on the *exact set* of channels allocated to user i, but only on the *size* of that set (k). This allows us to set up the following graph formulation of the throughput maximization problem in the high SINR regime.

In Figure 9.7, the L nodes representing the users are split up into M subnodes; one for each of the channels. The channels are represented

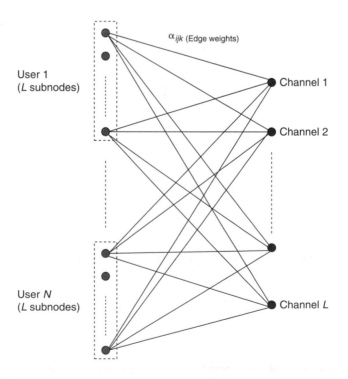

FIGURE 9.7
The constructed graph $\tilde{G}$.

separately using M nodes, as usual. All possible edges between the user subnodes and channels are drawn, with edge weights computed using Equation 9.42. Given this construction, note that a matching in the constructed bipartite graph $\tilde{G}$ corresponds to a polymatching in the original graph. Further, it can be verified that the edge-weights exhibit a decreasing property in k, that is, $\alpha_{ijk} > \alpha_{ij(k+1)}$ for any $k \geq 1$. If (i, j, k) denotes the edge between the kth subnode of user i and the channel j, then the decreasing property of the edge-weights imply that a *maximum weight matching* [3] (with α_{ijk} as the edge-weights) in $\tilde{G}$ will prefer edges that correspond to a lower k, for the same i and j. Thus, in a maximum weight matching, for any user i, there will be a k_i such that subnodes $1, \ldots, k_i$, will be matched, and subnodes $k_i + 1, \ldots, M$, would not be matched. This line of argument can be extended further to show that a maximum weight matching in $\tilde{G}$ corresponds to the polymatching that maximizes the sum of user throughput, where the user throughputs are defined by Equation 9.41. Therefore, in the high SINR regime, the optimum channel assignment can be calculated by computing the maximum weight matching in the constructed bipartite graph $\tilde{G}$, the complexity of which is $O(L^3 M^3)$ using the classical Hungarian algorithm [10]. We refer to this algorithm as the HSO (high SINR optimal) algorithm.

9.4.2 Throughput Analysis in the Low SINR Regime

In the low SINR regime, we approximate the objective function as

$$\log \left(1 + \frac{p_{ij}}{n_{ij}} \right) \approx \frac{p_{ij}}{n_{ij}} \tag{9.43}$$

using the approximation $\log(1 + x) \approx x$ when $0 < x \ll 1$. Further, if we assume that all n_{ij} values are distinct, then for small enough SINR, each user will allocate all its power in a single channel—the one with the smallest n_{ij} among all channels assigned to the user. More precisely, if the n_{ij} values differ at least by ϵ, then for $P_i < \epsilon$, user i will allocate all of its power to the minimum-noise channel it is allocated for maximum throughput. (Note that this situation is the opposite of the high SINR case, where the user will typically use all channels assigned to the user.) Using this fact, and Equation 9.43, we see that the channel assignment policy for maximum throughput in the low SINR regime corresponds to a maximum weight matching in the complete bipartite graph of users and channels, with edge-weights $\beta_{ij} = P_i / n_{ij}$. This maximum weight matching can be computed in $O(\{\max(L, M)\}^3)$ time, using the Hungarian algorithm [10].

Note that, if the number of channels is more than the number of users, the matching algorithm will leave a number of channels unassigned to any user. In practice, P_i can be larger than the minimum difference in the noise levels, and thus leaving available channels unassigned can lead to considerable wastage of resources. Therefore, we run the matching iteratively, leaving out all channels assigned in previous iterations, until all channels are allocated. We refer to this algorithm as the LSO (low SINR optimal) algorithm.

9.4.3 Performance Evaluation

Figures 9.8 through 9.11 show the simulation results for four different system models, which consist of three users and six channels, three users and nine channels, four users and eight channels, five users and ten channels, respectively. In all simulations, we choose $\sqrt{n_{ij}}$ from Gaussian distribution $N(0, \sigma^2)$; thus we have $E(n_{ij}) = \sigma^2$. For each user i, the maximum power P_i is chosen from the uniform distribution $U(0.5, 1.5)$; thus $E(P_i) = 1$. In the simulations, σ^2 is changed (keeping P_i fixed) to generate a wide range of SINR environments. For the same value of σ^2, simulations are run several times; the performance numbers shown in the figures across different SINR values correspond to the average performance for that SINR. In the figures, the x-axis corresponds to the SINR, plotted in a semilog scale. The y-axis corresponds to the ratio of the average throughput attained by an algorithm/heuristic and the maximum throughput attainable (solved by complete enumeration over all possible polymatchings). Note that the curves for max (HSO, LSO) in the figures show the best performance among the HSO and LSO algorithms.

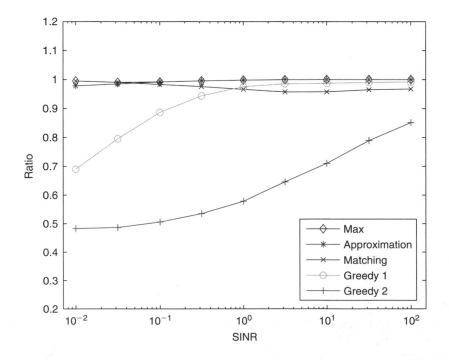

FIGURE 9.8
Simulation results for three users and six channels.

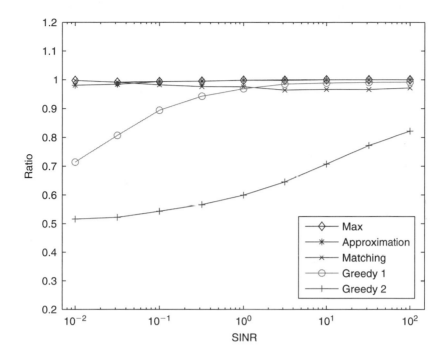

FIGURE 9.9
Simulation results for three users and nine channels.

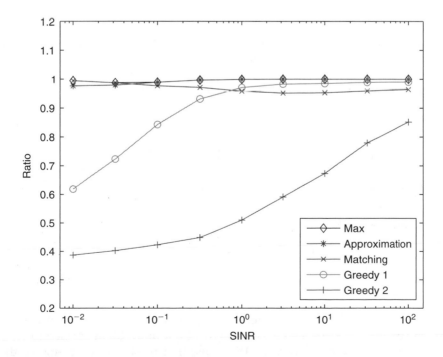

FIGURE 9.10
Simulation results for four users and eight channels.

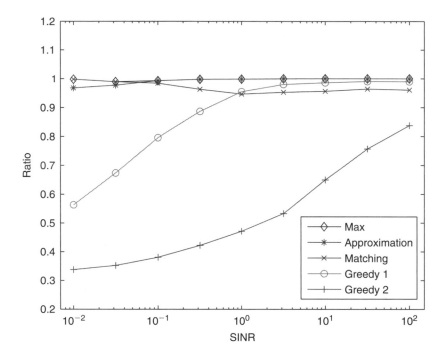

FIGURE 9.11
Simulation results for five users and ten channels.

From the figures, we see that the HSO algorithm achieves the optimal channel assignment (performance ratio is 1) under high SINR. In fact, the performance ratio of HSO is almost optimal when SINR is close to unity or higher. The figures also show that the LSO algorithm performs near optimally when SINR is low; its performance worsens as SINR increases, as expected.

We observe that the better of the HSO and LSO algorithms performs optimally over the entire range of SINR considered. This performance is also considerably better than that of the incremental heuristics. In practice, therefore, we can run both the HSO and LSO algorithms, and pick the better solution; this would result in near-optimal performance, no matter what the SINR value is, at only a small computation cost.

9.5 Summary and Open Problems

9.5.1 Summary

In this chapter, we have presented linear programming-based formulations for the demand constrained maximum throughput problem applied to IEEE 802.16-based wireless networks. We prove that the discrete version of the problem is NP-hard in general. We present an algorithm to find the maximum throughput, based on ideas from mixed covering and packing LPs. Owing to the dependence of the runtime of the algorithm on the best achievable data rate on any subchannel in the system, we also present a heuristic based on an interpretation as a generalized concurrent flow problem. The heuristic closely tracks the optimal value in numerical experiments. We also present the combined power and subchannel allocation problem for a single slot and present near-optimal algorithms for extremal SINR regimes.

9.5.2 Open Problems

There are some interesting questions yet to be answered. The question of how well the solution to the LP approximates the discrete version of the problem, which is NP-hard, is open. Also of interest are algorithms and heuristics for the online version of the throughput maximization problem, and efficient algorithms for the joint power and slot allocation formulation under QoS constraints. While initial directions have been presented in this chapter, their detailed analysis and study needs work. In conclusion, the area of frame-based, multitone, QoS constrained scheduling, and resource allocation are relatively new and have direct applications in next generation wireless networks.

References

1. H. Yaghoobi, *Scalable OFDMA Physical Layer in IEEE 802.16 WirelessMAN*, Intel Technology Journal, Vol. 8, Issue 3, pp. 201–212, 2004.
2. Draft IEEE Standard for Local and Metropolitan Area Networks, Part 16: Air Interface for Fixed Broadband Wireless Access Systems.
3. T. H. Cormen, C. E. Leiserson, and R. L. Rivest, *Introduction to Algorithms*, McGraw-Hill, New York, 1990.
4. L. Fleischer and K. Wayne, *Fast and Simple Approximation Schemes for Generalized Flow*, Mathematical Programming, Vol. 91, No. 2, pp. 215–238, 2002.
5. G. Kulkarni, S. Adlakha, and M. Srivastava, *Subcarrier Allocation and Bit Loading Algorithms for OFDMA Based Wireless Networks*, IEEE Transactions on Mobile Computing, Vol. 04, No. 6, pp. 652–662, November 2005.
6. N. Young, *Sequential and Parallel Algorithms for Mixed Covering and Packing*, Proceedings of Foundations of Computer Science 2001, p. 538.
7. M. Garey and D. Johnson, *Computers and Intractability*, W. H. Freeman, San Francisco, CA, 1979.
8. P. Crescenzi and V. Kann (Editors), *A Compendium of NP Optimization Problems*, available online at http://www.nada.kth.se/viggo/problemlist/compendium.html.
9. L. Fleischer and K. Wayne, *Faster Approximation Algorithms for Generalized Network Flow*, Proceedings of the ACM/SIAM Symposium on Discrete Algorithms, 1999.
10. H. W. Kuhn, *The Hungarian Method for the Assignment* problem, Naval Research Logistic Quarterly, Vol. 2, pp. 83–97, 1955.

10

Resource Allocation and Admission Control Using Fuzzy Logic for OFDMA-Based IEEE 802.16 Broadband Wireless Networks

Dusit Niyato and Ekram Hossain

CONTENTS

10.1   Introduction

IEEE 802.16 standard [1] and its evolutions (i.e., 802.16a, 802.16-2004, 802.16e, 802.16g) will provide high-bandwidth services with an array of multimedia features in the next generation wireless networks. Also, known as the WiMAX (worldwide interoperability for microwave access), IEEE 802.16 standard has been developed to provide broadband wireless access (BWA) with a flexible QoS framework. WiMAX networks can operate either on 10–66 or 2–11 GHz (IEEE 802.16a) band and support data rate in the range of 32–130 Mbps, depending on the bandwidth of operation as well as the modulation and coding schemes used. While in the 10–66 GHz band (i.e., with WirelessMAN-SC air interface), the signal propagation between a base station (BS) and a subscriber station (SS) or mobile must be line-of-sight, IEEE 802.16a operating in the 2–11 GHz band supports nonline-of-sight communication. One of the air interfaces for 802.16a, namely, WirelessMAN-OFDMA, is based on orthogonal frequency division multiple access (OFDMA) where the entire bandwidth is divided into subchannels, which are dynamically allocated among the different connections. Even though the physical layer and medium access control (MAC) protocols are well defined in the standard, the resource allocation and admission control are left open on purpose for innovations by individual equipment vendors.

The QoS framework defined in IEEE 802.16 considers four types of services: unsolicited grant service (UGS), real-time polling service (rtPS), nonreal-time polling service (nrtPS), and best-effort (BE) service. The UGS and the BE services are for constant bit rate QoS-sensitive traffic and for QoS-insensitive traffic, respectively. The polling service is intended to support both real-time and nonreal-time variable bit rate traffic. For nrtPS, only a certain level of throughput needs to be ensured, while for rtPS a strict delay requirement must be satisfied. Therefore, efficient allocation of radio resources for polling service is a critical issue in WiMAX-based BWA.

Different components of the radio resource management protocol such as traffic scheduling and admission control need to be designed for WiMAX networks such that the required QoS performances are guaranteed for the subscribers, as well as the resource utilization is maximized. Soft computing techniques (e.g., based on fuzzy logic, genetic algorithm) are effective approaches for designing radio resource management protocols to satisfy the QoS requirements of the users while maximizing the revenue of the system operators [2,3]. In this article, we apply fuzzy logic to solve the problem of radio resource management and admission control for WiMAX-based broadband wireless access networks using OFDMA air interface.

The motivation of using fuzzy logic for admission control in wireless networks comes from the fact that the system parameters (e.g., channel quality measurements, mobility estimation parameters, traffic source parameters) are often very imprecise and cannot be estimated very accurately. It is difficult to design efficient resource allocation and admission control policies based on these imprecise information. Applying the traditional admission control algorithms, therefore, might not be able to achieve desired system performance [2]. Again, since fuzzy logic control is based on simple rule-based inference system, efficient solutions can be achieved with relatively low complexity.

Owing to the simplicity in modeling and the ability to generate predictable output from uncertain inputs, fuzzy logic is a promising technique for resource allocation and admission control for IEEE 802.16-based broadband wireless access. To design the fuzzy controller, we use an approach shown in Figure 10.1 [4]. In particular, for a WiMAX system we first develop a queueing model based on discrete-time Markov chain (DTMC) to analyze packet-level QoS performances (e.g., average delay) for OFDMA transmission under adaptive modulation and coding (AMC). By using this queueing model in an off-line manner, the impacts of the traffic source parameters (i.e., arrival rate, peak rate, and probability of peak rate) and channel quality (i.e., average signal-to-noise ratio [SNR]) on the radio-link-level performances (i.e., average queue length, packet-dropping probability, throughput, and delay) can be investigated. Then, from the queueing model, we establish a set of rules for fuzzy control, which is used in an online manner for radio resource management and admission control. Since the admission control decision is made based on the amount of required resources to satisfy the QoS requirements of both the ongoing connections and the incoming connection, the results obtained from the queueing model are useful to design the resource allocator in the proposed fuzzy logic control system. The performance of the proposed admission control scheme is analyzed by simulations and compared with that of some of the traditional schemes (e.g., static and adaptive admission control schemes).

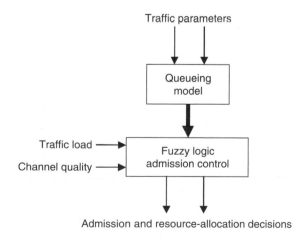

FIGURE 10.1
Queueing model and fuzzy logic controller.

The rest of this chapter is organized as follows. Section 10.2 reviews some of the related works on the application of fuzzy logic for radio resource management in wireless networks. The basics of fuzzy logic and the different components of a fuzzy logic controller are described in Section 10.3. Section 10.4 describes the WiMAX system model considered in this article. The queueing formulation for the system model is presented in Section 10.5. Section 10.6 presents the admission control scheme based on fuzzy logic control. The numerical and simulation results are presented in Section 10.7. Section 10.8 states the conclusions.

10.2   Related Work

In a wireless network, radio resource allocation and admission control methods significantly impact the packet-level QoS performance (e.g., delay, throughput, loss). While the resource-allocation policies ensure that the required amount of radio resources are allocated to the different connections so that the QoS requirements can be guaranteed, an admission control scheme limits the number of ongoing connections so that the network is not overloaded owing to the admission of too many users. In the literature, many of the resource-allocation and admission-control methods are based on optimization models [5–7] to maximize or minimize a defined objective function (e.g., maximizing system utilization, user utility, or minimizing packet delay and loss). However, to achieve the optimal solution by using an optimization technique we require complete, correct, and precise information of the system. Also, in most of the cases a closed form solution is not available, and the algorithm to solve the optimization formulation is computation intensive. Therefore, this approach may not be feasible or efficient for online operation of the protocol in a practical wireless system.

Alternatively, fuzzy logic can be applied to solve the problem of resource management and connection admission control in wireless systems. A fuzzy logic controller can be designed to achieve a near optimal solution and it also incurs much less computational complexity. Therefore, it might be more suitable for online operation of the protocol in a wireless system, in which solution is required online to improve the QoS performance of the system.

For circuit-switched wireless networks, a fuzzy logic-based radio resource controller was developed for resource reservation [8]. This fuzzy logic controller was designed to differentiate real-time traffic from nonreal-time traffic. The effective transmission rate was estimated to optimally allocate the available time slots to nonreal-time traffic, so that the highest system utilization can be achieved while guaranteeing the QoS performances of real-time traffic.

Fuzzy logic was applied for call admission control and radio resource management in code division multiple access (CDMA)-based wireless networks [2–11]. In Ref. 2, fuzzy logic was used to estimate the effective bandwidth for a call and the mobility information for the mobile. Then, the admission control decision (i.e., accepting or rejecting an incoming call) was made based on these estimations and the amount of available resource. From a simulation-based performance study, the authors concluded that the fuzzy logic-based admission control could satisfy the QoS requirements in terms of outage probability, new call-blocking, and handoff call-dropping probabilities while maximizing the resource utilization. With a similar design philosophy, in Ref. 4, fuzzy logic was used to estimate interference power from an incoming call. Also, a pipeline recurrent neural network (PRNN) was used to predict the system interference in the next time period. Then, the fuzzy call admission controller used these estimations, as well as the QoS performance requirements for each of the traffic types and the channel quality information to decide whether an incoming call can be accepted or not. The proposed estimation and prediction mechanisms were observed to improve the system capacity significantly. However, the admission controller considered only bit-level (e.g., bit-error rate) and connection-level performances (e.g., call-blocking probability) and ignored the packet-level performances (e.g., delay and throughput). In Ref. 9, fuzzy logic was used for radio resource management in time-division CDMA (TD-CDMA) networks with space division multiple access (SDMA) scheduling. Real-time (e.g., voice and video) and BE traffic (e.g., data) were considered in the system model. Fuzzy logic controller was designed to optimize both call-blocking and call-dropping probabilities for real-time calls and packet delay for data traffic.

In Refs. 10 and 11, fuzzy logic and neural system were used for radio resource management among the different access networks. With three different radio access technologies (RATs), namely, universal mobile telecommunications system (UMTS), GSM/EDGE radio access network (GERAND), and wireless local area network (WLAN), fuzzy logic was used to estimate signal strength, radio resource availability, and mobile speed to allocate optimal bandwidth and to select the best RAT for the mobile. Also, a reinforcement learning based on neural network was used to adjust the parameters of fuzzy logic control (i.e., rules and membership functions) according to the changing environment. The network performance was studied in terms of call-blocking and call-dropping probabilities.

Fuzzy logic was used to solve the problem of admission control in IEEE 802.11 WLANs. In Ref. 12, fuzzy logic was used to estimate network congestion in terms of load and error rate, and connection acceptance decision was made based on these information. In Ref. 13, the model was extended to support differentiated QoS so that the throughput requirements for the connections in different priority classes are satisfied.

Fuzzy logic was used for radio resource management during handoff in cellular wireless networks. In Ref. 14, fuzzy logic was used to reserve radio bandwidth to minimize handoff call-dropping probability. For CDMA networks, a soft handoff algorithm based on fuzzy logic was proposed in Ref. 15 to provide fair distribution of network resources and minimize call-outage probability. The inputs of this fuzzy logic controller were the number of BSs and the number of free channels in each of the BS. Also, fuzzy logic was used to estimate parameters such as signal strength, mobile speed, and traffic load in each cell [14–26] to assist handoff initiation.

10.3 Fuzzy Logic

10.3.1 Introduction

The theory of fuzzy logic was developed with an objective to present approximate knowledge, which may not be suitably expressed by conventional crisp method (i.e., bivalent set theory). In the conventional method, *truth* and *false* are represented by values 1 and 0, respectively. By using membership functions, fuzzy logic extends and generalizes these truth and false values to any value between 0 and 1. A membership function indicates the degree by which the value belongs to a fuzzy set.

Fuzzy logic can be used to make a decision by using incomplete, approximate, and vague information. Therefore, it is suitable for a complex system for which it is difficult to compute/express the parameters precisely (e.g., by using a mathematical model). Fuzzy logic also enables us to make inferences by using the approximate information to decide on an appropriate action for given input. In short, instead of using complicated mathematical formulations, fuzzy logic uses human-understandable fuzzy sets and inference rules (e.g., IF, THEN, ELSE, AND, OR, NOT) to obtain the solution that satisfies the desired system objectives.

Fuzzy logic is computation-friendly and the complexity involved in problem solving is low. Therefore, fuzzy logic is suitable for real-time applications in which the response time is critical to the system.

10.3.2 Fuzzy Set

The fuzzy set theory is similar to traditional bivalent set theory. However, a fuzzy set has no clear or crisp boundaries. Also, since the set is represented by membership functions, it is possible that one element belongs to more than one set (with different degree). For example, say, there are two sets to represent room temperature (e.g., "hot" and "cold"). For a measured temperature (which could be inaccurate) of 28°C, it will be either "hot" or "cold" (e.g., if the threshold for "hot" is 30°C, this measured temperature is precisely "cold"). However, in the fuzzy logic, this measured temperature could be in set "hot" with a membership value of 70% and in set "cold" with a membership value of 20% (i.e., not quite cold, relatively hot). Note that, here temperature is called a fuzzy variable, while "hot" or "cold" are called linguistic variables.

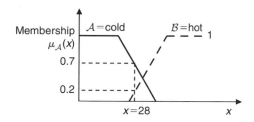

FIGURE 10.2
Membership functions for room temperature.

The terminologies commonly used in fuzzy set are as follows:

- *Universe of discourse* is the range of all possible values for an input to a fuzzy logic system (e.g., 0°C–40°C for room temperature).
- *Fuzzy set* is any set that allows its members to have different degree of membership (through membership function) in the interval [0,1] (e.g., "hot" and "cold").
- *Membership function* returns the degree of membership of a particular element within the set.

The membership function assigns a value in the interval [0,1] to a fuzzy variable. This membership function represents the possibility function (i.e., not probability function) of an element in a particular fuzzy set. The membership function can be expressed as $\mu_A(a)$, where A is a fuzzy set and a an element in the universe of discourse. An example of membership function is shown in Figure 10.2, where fuzzy sets A and B are for "cold" and "hot," respectively.

10.3.3 Fuzzy Operation

The operations on fuzzy set are similar to those in bivalent set theory (i.e., NOT, OR, AND).

- *NOT* denotes the complement of a fuzzy set A (i.e., A'), and the corresponding operation on membership function is given by $\mu_{A'}(a) = 1 - \mu_A(a)$.
- *OR* denotes the union of fuzzy sets $A \cup B$, and the corresponding operation on membership function is given by $\mu_{A \cup B}(a) = \max(\mu_A(a), \mu_B(a))$. Since this OR operation is a union of multiple fuzzy sets, membership function of this operation is the largest membership from all fuzzy sets.
- *AND* denotes the intersection of fuzzy set $A \cap B$, and the corresponding operation on membership function is given by $\mu_{A \cap B}(a) = \min(\mu_A(a), \mu_B(a))$. Since this AND operation is the

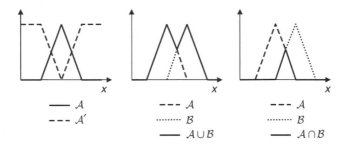

FIGURE 10.3
Operations on fuzzy sets.

intersection of multiple fuzzy sets, the membership function is the lowest membership from all fuzzy sets.

The graphical presentations of these operations are shown in Figure 10.3.

10.3.4 Fuzzy Rule

Knowledge in fuzzy logic is represented in the form of linguistic rules. A rule is based on cause and effect, which is in IF–THEN format (i.e., implication). The knowledge base is composed of several fuzzy rules. To determine the outcome (or decision), these rules are evaluated and the outcomes are aggregated to the final solution. If A and C denote fuzzy sets, the rule representing IF A THEN C can be expressed as $A \rightarrow C$. In this rule, A is called cause, condition, or antecedent of the rule, while C is called effect, action, or consequent of the fuzzy rule. For this IF–THEN rule, the membership function of the outcome c for given input a can be obtained in many different approaches as follows [35]:

- *Larsen implication:* $\mu_{A \rightarrow C}(a, c) = \mu_A(a)\mu_C(c)$
- *Mamdani implication:* $\mu_{A \rightarrow C}(a, c) = \min(\mu_A(a), \mu_C(c))$
- *Zadeh implication:* $\mu_{A \rightarrow C}(a, c) = \max[\min(\mu_A(a), \mu_C(c)), 1 - \mu_A(a)]$
- *Dienes–Rescher implication:* $\mu_{A \rightarrow C}(a, c) = \max(1 - \mu_A(a), \mu_C(c))$
- *Lukasiewicz implication:* $\mu_{A \rightarrow C}(a, c) = \min(1, 1 - \mu_A(a) + \mu_C(c))$

Note that, Mamdani implication is the most commonly used as it can provide correct and robust outcome.

10.3.5 Fuzzy Logic Control

A fuzzy logic control system provides a simple way to obtain solution to a problem based on imprecise, noisy, and incomplete input information. In general, there are three major components in a fuzzy logic control system: fuzzifier, fuzzy logic processor, and defuzzifier (Figure 10.4).

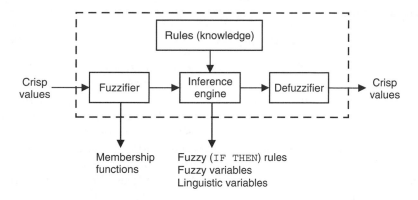

FIGURE 10.4
Fuzzification, inference engine, and defuzzification.

While the fuzzifier is used to map the crisp inputs into fuzzy set, the fuzzy logic processor implements an inference engine to obtain the solution based on predefined sets of rules. Then, the defuzzifier is applied to transform the solution to the crisp output.

In the fuzzification process, input values are fuzzified to determine the membership functions. Then, these fuzzified inputs are used by the inference rules to determine an outcome or decision. For example, let $\mathcal{A}_i$ and $\mathcal{B}_i$ denote the fuzzified inputs, and $\mathcal{C}_i$ denote an output. A set of rules can be defined as follows:

> Rule 1: $\mathcal{A}_1$ AND $\mathcal{B}_1 \rightarrow \mathcal{C}_1$
>
> Rule 2: $\mathcal{A}_2$ AND $\mathcal{B}_2 \rightarrow \mathcal{C}_2$
>
> Rule n: $\mathcal{A}_n$ AND $\mathcal{B}_n \rightarrow \mathcal{C}_n$

Since these rules are related through IF–THEN implication, the membership function of an output of a particular rule can be expressed as $\mu_{R_i}(a, b, c) = \min(\mu_{\mathcal{A}}(a), \mu_{\mathcal{B}}(b), \mu_C(c))$. Then, the outcomes of all rules are combined using maximum function of each rule as follows:

$$\mu_R(a, b, c) = \max_i \min(\mu_{\mathcal{A}_i}(a), \mu_{\mathcal{B}_i}(b), \mu_{C_i}(c))$$

However, this outcome from all rules determines the membership function not the crisp value. Therefore, we need the defuzzification process to obtain the final output of the controller. The most widely used method is centroid, in which the output is determined from a center of gravity of the membership function from the outcome of the set of rules. Let $\mathcal{K} = \{c | \mu_C(c) > 0\}$ denote a set of outputs c with membership value larger than zero. Then, the defuzzified output can be obtained from $\hat{c} = \int_{c \in \mathcal{K}} c \mu_C(c) \, dc / \int_{c \in \mathcal{K}} \mu_C(c) \, dc$, if set $\mathcal{K}$ is continuous and $\hat{c} = \sum_{c \in \mathcal{K}} c \mu_C(c) / \sum_{c \in \mathcal{K}} \mu_C(c)$, if $\mathcal{K}$ is a discrete set.

10.4 WiMAX System Model

We consider a downlink communication scenario between a BS and SSs operating in TDD–OFDMA mode with C subchannels available to serve multiple connections. The frame structure is shown in Figure 10.5 [27], in which the frame is divided into downlink and uplink subframes. Each subframe is composed of multiple bursts and each burst is used for transmission of protocol data units (PDUs) corresponding to one connection. A single burst can carry several PDUs on multiple subchannels, and a subchannel can be shared by several bursts. Adaptive modulation and coding is used to adjust the transmission rate in each subchannel dynamically according to the channel quality. The modulation levels and coding rates, information bits per symbol, and required SNR for IEEE 802.16 air interface are shown in Table 10.1. With basic modulation and coding scheme (i.e., *rate ID* = 0), one subchannel can transmit L PDUs (e.g., $L = 3$ in Figure 10.5). Therefore, the total PDU transmission rate

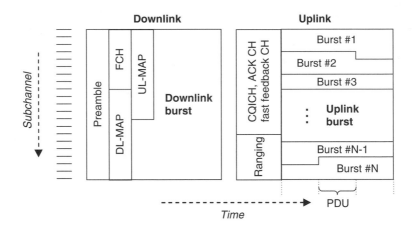

FIGURE 10.5
Frame structure of IEEE 802.16 with TDD–OFDMA mode.

TABLE 10.1

Modulation and Coding Schemes

Rate ID	Modulation Level (Coding)	Information (Bits/Symbol)	Required SNR
0	BPSK (1/2)	0.5	6.4
1	QPSK (1/2)	1	9.4
2	QPSK (3/4)	1.5	11.2
3	16QAM (1/2)	2	16.4
4	16QAM (3/4)	3	18.2
5	64QAM (2/3)	4	22.7
6	64QAM (3/4)	4.5	24.4

depends on the number of allocated subchannels and *rate ID* used in each subchannel.

We assume that the subchannel condition remains stationary over a frame interval (≤ 2 ms), and all the PDUs transmitted in the same subchannel during one frame period use the same *rate ID*.

10.5 Queueing Formulation

The delay incurred for successful transmission of a packet across a wireless link depends largely on the radio link-level queue management and error-control methods. The problem of analyzing radio link-level queueing under wireless packet-transmission was addressed in the literature. In Ref. 28, a Markov-based model was presented to analyze the radio link-level PDU dropping process under automatic repeat request (ARQ)-based error control. In Ref. 29, a cross-layer analytical model to derive PDU loss rate, average throughput, and delay under AMC was presented. However, these queueing models considered transmission of a single user in a single channel only. Queueing models were proposed for WiMAX system. For example, queueing models for performance analysis at WiMAX BS and SS were proposed in Refs. 30 and 31, respectively. We investigate packet-level performance by formulating a queueing model specifically for OFDMA wireless transmission. We also consider the burstiness of traffic arrival by modeling the traffic source by a Markov modulated Poisson process (MMPP). From this queueing model, various packet-level performance measures can be obtained, which can be used to establish the rules for fuzzy controller for radio resource management and connection admission control.

10.5.1 Traffic Source and Arrival Probability Matrix

To capture the peak arrival rate for a traffic source, we use an MMPP model, which is a general traffic model for multimedia traffic as well as Internet traffic [32]. With MMPP, the PDU arrival rate λ_s is determined by the phase s of the Markov chain, and the total number of phases is S (i.e., $s = 1, 2, \ldots, S$). An MMPP can be represented by matrices $\mathbf{M}$ and $\mathbf{\Lambda}$, in which the former is the transition probability matrix of the modulating Markov chain, and the latter is the matrix corresponding to the Poisson arrival rates. Since we consider traffic source with normal rate and peak rate only, the number of phases of the MMPP model is two (i.e., $S = 2$), and the matrices are defined as follows:

$$\mathbf{M} = \begin{bmatrix} 1 - P_{\text{peak}}, & P_{\text{peak}} \\ 1 - P_{\text{peak}}, & P_{\text{peak}} \end{bmatrix}, \mathbf{\Lambda} = \begin{bmatrix} \lambda & \\ & \lambda_p \end{bmatrix} \qquad (10.1)$$

where P_{peak} is the probability of peak arrival rate, λ is the normal PDU arrival rate, and λ_p is the peak rate.

Discrete-time MMPP (dMMPP) [33] is equivalent to MMPP in the continuous time. In this case, the rate matrix Λ is represented by diagonal probability matrix Λ_a when the number of PDUs arriving in one frame is a. Each element of Λ_a can be obtained as follows:

$$\Lambda_a = \begin{bmatrix} f_a(\lambda) \\ & f_a(\lambda_p) \end{bmatrix}, \dots, \Lambda_A = \begin{bmatrix} F_A(\lambda) \\ & F_A(\lambda_p) \end{bmatrix} \qquad (10.2)$$

where the probability that a PDUs arrive during time interval t (i.e., frame period) with mean rate λ is given by

$$f_a(\lambda) = \frac{e^{-\lambda t}(\lambda t)^a}{a!}$$

where $a \in \{0, 1, \dots, A\}$ and A is the maximum batch size for PDU arrival. The complementary cumulative probability mass function for this arrival process is given by

$$F_a(\lambda) = \sum_{j=a}^{\infty} f_j(\lambda, t)$$

10.5.2   Transmission in the Subchannels

We consider a Nakagami-m channel model for each subchannel, in which the channel quality is determined by the instantaneous received SNRγ. With adaptive modulation, SNR is divided into $N + 1$ nonoverlapping intervals (i.e., $N = 7$ in IEEE 802.16) by thresholds Γ_n ($n \in \{0, 1, \dots, N\}$), where $\Gamma_0 < \Gamma_1 < \cdots < \Gamma_{N+1} = \infty$. The subchannel is said to be in state n (i.e., *rate ID* $= n$ will be used), if $\Gamma_n \le \gamma < \Gamma_{n+1}$. To avoid possible transmission error, no PDU is transmitted when $\gamma < \Gamma_0$. Note that, these thresholds correspond to the required SNR specified in the IEEE 802.16 standard, that is, $\Gamma_0 = 6.4, \Gamma_1 = 9.4, \dots, \Gamma_N = 24.4$ (as shown in Table 10.1). From Nakagami-m distribution, the probability of using *rate ID* $= n$ (i.e., $Pr(n)$) can be obtained as follows [29]:

$$Pr(n) = \frac{\Gamma(m, m\Gamma_n/\bar{\gamma}) - \Gamma(m, m\Gamma_{n+1}/\bar{\gamma})}{\Gamma(m)} \qquad (10.3)$$

where $\bar{\gamma}$ is the average SNR, m the Nakagami fading parameter ($m \ge 0.5$), $\Gamma(m)$ the gamma function, and $\Gamma(m, \gamma)$ the complementary incomplete gamma function.

Let $\mathcal{C}$ denote the set of allocated subchannels for a particular connection. We can define row matrix $\mathbf{D}_c$ whose elements d_{k+1} correspond to the probability of transmitting k PDUs in one frame on one subchannel c ($c \in \mathcal{C}$) as follows:

$$\mathbf{D}_c = \begin{bmatrix} d_0 \cdots d_k \cdots d_{9 \times L} \end{bmatrix} \qquad (10.4)$$

where

$$d_{(I_n \times 2 \times L)} = Pr(n) \qquad (10.5)$$

where I_n is the number of transmitted bits per symbol corresponding to *rate ID* $= n$ and $d_0 = 1 - \sum_{k=1}^{9 \times L} d_k$. This matrix $\mathbf{D}_c$ has size $1 \times (9 \times L) + 1$. For the subchannel that the connection shares with another connection (i.e., L_s out of L PDUs in one frame in that shared channel), this matrix becomes

$$\mathbf{D}_c = \begin{bmatrix} d_0 \cdots d_k \cdots d_{9 \times L_s} \end{bmatrix} \qquad (10.6)$$

Similarly, $d_{(I_n \times 2 \times L_s)} = Pr(n)$. Note that, we can calculate the total number of allocated subchannels c_i based on *Rate ID* $= 0$ for connection i from

$$c_i = \text{ele}(\mathcal{C}) + \frac{L_s}{L} \qquad (10.7)$$

where function $\text{ele}(\mathcal{C})$ gives the number of elements in set $\mathcal{C}$.

The matrix for *pmf* of total PDU transmission rate can be obtained by convoluting matrices $\mathbf{D}_c$ as follows:

$$\mathbf{D} = \odot \, \forall c \in \mathcal{C} \, \mathbf{D}_c$$

where $\mathbf{a} \odot \mathbf{b}$ denotes discrete convolution [34] between matrices $\mathbf{a}$ and $\mathbf{b}$. Note that, matrix $\mathbf{D}$ has size $1 \times U + 1$, where $U = (9 \times \lfloor c_i \rfloor \times L) + (9 \times L_s) + 1$ indicates the maximum number of PDUs that can be transmitted in one frame.

The total PDU transmission rate can be obtained as follows:

$$\alpha - \sum_{k=1}^{U} k \times [\mathbf{D}]_{k+1} \qquad (10.8)$$

10.5.3   State Space and Transition Matrix

For each connection, a separate queue with size X PDUs is used for buffering data from the corresponding application. The state of the queue (i.e., the number of PDUs in the queue and phase of arrival) is observed at the beginning of each frame. A PDU arriving in frame f will not be transmitted until the next frame $f + 1$ at the earliest. The state space of a queue can be defined as follows:

$$\Phi = \{(\mathcal{X}, \mathcal{M}), \, 0 \leq \mathcal{X} \leq X, \mathcal{M} \in \{1, 2\}\}$$

where $\mathcal{X}$ and $\mathcal{M}$ indicate the number of PDUs in the queue and the phase of the MMPP arrival process, respectively. The transition matrix $\mathbf{P}$ of the queue can be expressed as in Equation 10.9. The rows of matrix $\mathbf{P}$ represent the number of PDUs in the queue, and element $\mathbf{p}_{x,x'}$ inside this matrix denotes the probability matrix for the case when the number of PDUs in the queue changes from x in the current frame to x' in the next frame.

$$
P = \begin{bmatrix}
p_{0,0} & \cdots & p_{0,A} & & & & \\
\vdots & \ddots & \ddots & \ddots & & & \\
p_{U,0} & \cdots & p_{U,U} & \cdots & p_{U,U+A} & & \\
& \ddots & \ddots & \ddots & \ddots & \ddots & \\
& p_{x,x-U} & \cdots & p_{x,x} & \cdots & p_{x,x+A} & \\
& & \ddots & \ddots & \ddots & \ddots & \\
& & & p_{X,X-U} & \cdots & p_{X,X}
\end{bmatrix}.
\qquad (10.9)
$$

Since in one frame several PDUs can arrive and be transmitted, this matrix **P** is divided into three parts. The first part, from row 0 to $U - 1$, indicates the case that the maximum total transmission rate is greater than the number of PDUs in the queue and none of the incoming PDUs is dropped. The second part, from row U to $X - A$, represents the case in which the maximum PDU transmission rate is equal to or less than the number of PDUs in the queue and none of the incoming PDUs is dropped. Since the size of queue is finite, some of the arriving PDUs will be dropped because of the lack of buffer space. The third part, from row $X - A + 1$ to X, indicates the case that some of the incoming PDUs are dropped because of the lack of space in the queue. Let $\mathbf{D}^{(x)}$ denote transmission probability when there are x PDUs in the queue and it can be obtained from $\mathbf{D}^{(x)} = \begin{bmatrix} d_0 & \cdots & d_{U'} \end{bmatrix}$, where $U' = \min(x, U)$ and

$$
d_{U'} = \begin{cases} d_U, & U' = U \\ \sum_{k=x}^{U} d_k, & U' = x \end{cases}
\qquad (10.10)
$$

Note that, since the maximum total PDU transmission rate can be greater than the number of PDUs in the queue, the maximum number of transmitted PDUs cannot be larger than the number of PDUs in the queue.

The elements in the first and the second part of matrix **P** can be obtained as follows:

$$
p_{x,x-u} = \mathbf{M} \left\{ \sum_{k-j=u} \mathbf{\Lambda}_j \times \left(\left[\mathbf{D}^{(x)} \right]_{k+1} \mathbf{I}_2 \right) \right\}
\qquad (10.11)
$$

$$
p_{x,x+v} = \mathbf{M} \left\{ \sum_{j-k=v} \mathbf{\Lambda}_j \times \left(\left[\mathbf{D}^{(x)} \right]_{k+1} \mathbf{I}_2 \right) \right\}
\qquad (10.12)
$$

$$
p_{x,x} = \mathbf{M} \left\{ \sum_{k=j} \mathbf{\Lambda}_j \times \left(\left[\mathbf{D}^{(x)} \right]_{k+1} \mathbf{I}_2 \right) \right\}
\qquad (10.13)
$$

for $u = 1, \ldots, U', v = 1, \ldots, A$, where $k \in \{0, 1, 2, \ldots, U'\}$ and $j \in \{0, 1, 2, \ldots, A\}$ represent the number of departed PDUs and the number of arriving PDUs, respectively, and I_2 is an identity matrix of size 2×2.

Considering both the PDU arrival and the PDU departure events, Equations 10.11 through 10.13 represent the transition probability matrices for the cases when the number of PDUs in the queue decreases by u PDUs, increases by v PDUs, and does not change, respectively.

The third part of matrix $\mathbf{P}$ ($\{x = X - A + 1, X - A + 2, \ldots, X\}$) has to capture the PDU dropping effect. Therefore, for $x + v \geq X$, Equation 10.12 becomes

$$\mathbf{p}_{x,x+v} = \sum_{i=v}^{A} \hat{\mathbf{p}}_{x,x+i} \quad \text{for } x + v \geq X \tag{10.14}$$

where $\hat{\mathbf{p}}_{x,x'}$ is obtained considering that there is no packet drop. For $x = X$, Equation 10.13 becomes

$$\mathbf{p}_{x,x} = \hat{\mathbf{p}}_{x,x} + \sum_{i=1}^{A} \hat{\mathbf{p}}_{x,x+i} \tag{10.15}$$

Equations 10.14 and 10.15 indicate the case that the queue will be full if the number of incoming PDUs is greater than the available space in the queue.

10.5.4 QoS Measures

To obtain the performance measures, the steady-state probabilities for the queue would be required. Since the size of the queue is finite (i.e., $X < \infty$), the probability matrix π is obtained by solving the equations $\pi\mathbf{P} = \pi$ and $\pi\mathbf{1} = 1$, where $\mathbf{1}$ is a column matrix of ones. The matrix π contains the steady-state probabilities corresponding to the number of PDUs in the queue and the phases of MMPP traffic source. This matrix π can be decomposed to $\pi(x, s)$, which is the steady-state probability that there are x PDUs in queue and MMPP phase is s, as follows $\pi(x, s) = [\pi]_{2x+s}$. Here, $s = 1$ and 2 indicate that the traffic source transmits with the normal and the peak rate, respectively.

10.5.4.1 Average Number of PDUs in Queue

The average number of PDUs in the queue is obtained as follows:

$$\bar{x} = \sum_{x=1}^{X} x \left(\sum_{s=1}^{2} \pi(x, s) \right) \tag{10.16}$$

10.5.4.2 PDU Dropping Probability

It refers to the probability that an incoming PDU will be dropped owing to the unavailability of buffer space. It can be derived from the average number of dropped PDUs per frame following the method used in Ref. 28. Given that there are x PDUs in the queue and the number of PDUs in queue increases by v, the number of dropped PDUs is $v - (X - x)$ for $v > X - x$, and zero otherwise. The average number of dropped PDUs per frame is obtained as follows:

$$\bar{x}_{\mathrm{drop}} = \sum_{s=1}^{2} \sum_{x=0}^{X} \sum_{v=X-x+1}^{A} \pi(x,s) \left(\sum_{j=1}^{2} [\mathbf{p}_{x,x+v}]_{s,j} \right) (v - (X - x)) \qquad (10.17)$$

where the term $\sum_{j=1}^{2} [\mathbf{p}_{x,x+v}]_{s,j}$ in Equation 10.17 indicates the total probability that the number of PDUs in the queue increases by v at every arrival phase. Note that, we consider probability $\mathbf{p}_{x,x+v}$ rather than the probability of PDU arrival since we have to consider the PDU transmission in the same frame as well. After calculating the average number of dropped PDUs per frame, we can obtain the probability that an incoming PDU is dropped as follows:

$$P_{\mathrm{drop}} = \frac{\bar{x}_{\mathrm{drop}}}{\bar{\lambda}}$$

where $\bar{\lambda}$ is the average number of PDU arrivals per frame and it can be obtained from $\bar{\lambda} = \lambda(1 - P_{\mathrm{peak}}) + \lambda_p P_{\mathrm{peak}}$.

10.5.4.3   Queue Throughput

It measures the number of PDUs transmitted in one frame and can be obtained from $\eta = \lambda(1 - P_{\mathrm{drop}})$.

10.5.4.4   Average Delay

It is defined as the number of frames that a PDU waits in the queue since its arrival before it is transmitted. We use Little's law to obtain average delay as follows:

$$\bar{w} = \frac{\bar{x}}{\eta}$$

where η is the throughput (same as the effective arrival rate at the queue) and $\bar{x}$ the average number of PDUs in queue.

10.6   Fuzzy Logic Controller for Admission Control

We propose an open-loop fuzzy logic control system for admission control, and the major components in this system are shown in Figure 10.6.

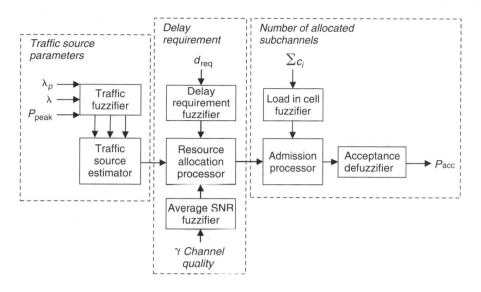

FIGURE 10.6
Block diagram of fuzzy admission control.

The admission control process works as follows. When a new connection is initiated, the corresponding mobile node informs the BS with the approximate traffic source parameters (i.e., normal rate λ, peak rate λ_p, and probability of peak rate P_{peak}) and the target delay requirement d_{req}. These inputs are fuzzified into fuzzy sets according to the corresponding membership functions. The traffic source estimator estimates traffic intensity as the output. Next, the BS measures and fuzzifies average SNR ($\bar{\gamma}$) corresponding to a new connection. This traffic intensity and channel quality information are used by the resource-allocation processor together with the user-specified delay requirement to obtain the number of subchannels to be assigned. The number of subchannels is bounded by c_{min} and c_{max} to ensure that the connection is allocated neither too large nor too small amount of transmission resource.

The number of allocated subchannels and the fuzzified amount of load in the cell $\sum_{i=1}^{N} c_i$, where N is the total number of ongoing connections, are used by the admission controller to decide whether an incoming connection can be accepted or not. Specifically, the output from this admission controller is defuzzified to acceptance probability P_{acc}, and the BS accepts the new connection based on this probability. The membership functions of inputs for traffic source estimator, resource allocation, and admission controllers are graphically shown in Figures 10.7 through 10.9, respectively, and Tables 10.2 through 10.4 show the corresponding inference rules. Note that, the rationale behind the establishment of these membership functions and the inference rules will be described in the next section. In each of these tables, the last

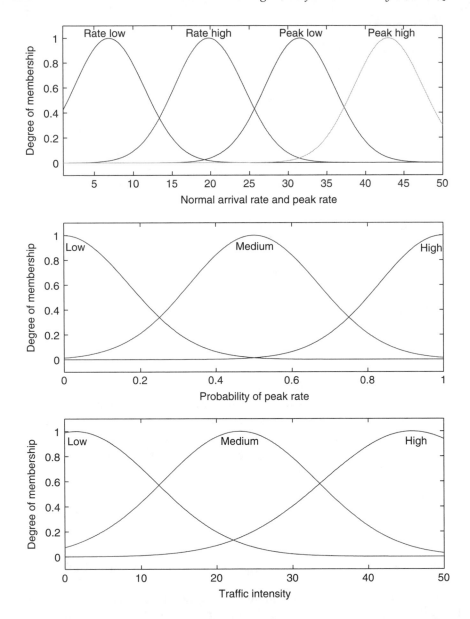

FIGURE 10.7
Membership functions corresponding to normal arrival rate, peak rate, and probability of peak rate.

column represents the *consequent*, and the first few columns correspond to the *antecedent*. The relationship between the antecedent and the consequent can be built through an IF–THEN statement and by using the AND operator between two antecedents. For example, the first rule in Table 10.2 is

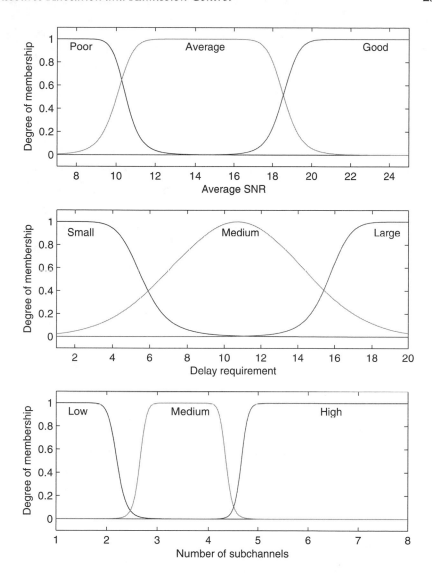

FIGURE 10.8
Membership functions corresponding to traffic intensity, average SNR, delay requirements, and number of subchannels.

```
IF (normal rate is low) AND (peak rate is low) AND
(probability of peak rate is NOT high) THEN traffic
intensity is low
```

and the last rule is

```
IF (normal rate is high) AND (peak rate is high) THEN
(traffic intensity is high).
```

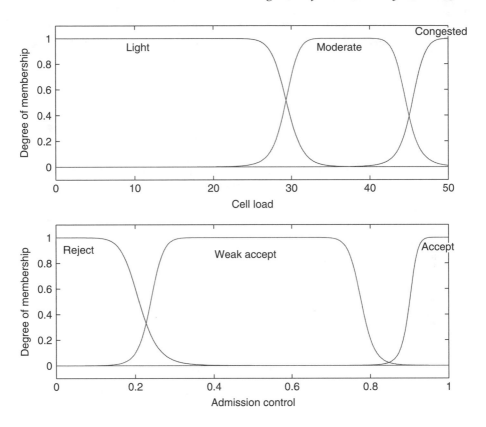

FIGURE 10.9
Membership functions corresponding to traffic load in a cell and the acceptance probability.

10.7 Performance Evaluation

10.7.1 Parameter Setting

We consider a BS with 50 subchannels (i.e., $C = 50$). Each subchannel has bandwidth of 160 kHz. The length of a subframe for downlink transmission is 1 ms, and therefore, the transmission rate in one subchannel with *rate ID* $= 0$ (i.e., BPSK modulation and coding rate is 1/2) is 80 Kbps. We assume that the size of each PDU is 80 bits and one subchannel can carry one PDU (i.e., $L = 1$) in one downlink subframe. Adaptive modulation and coding is performed independently in each subchannel to increase the transmission rate if the channel quality (i.e., average SNR) permits. Although we assume that the average SNR is the same for all the allocated subchannels to a particular connection, the instantaneous SNR, and consequently, the selected *rate ID* in

TABLE 10.2

Fuzzy Inference Rules for the Traffic Source Estimator

Normal Rate	Peak Rate	P_{peak}	Traffic Intensity
Low	Low	NOT high	Low
Low	Low	High	Medium
Low	High	Low	Medium
Low	High	NOT low	High
High	Low	NOT high	Medium
High	Low	High	High
High	High	–	High

TABLE 10.3

Fuzzy Inference Rules for the Resource-Allocation Processor

Traffic Intensity	Average SNR	Required Delay	Required Resource
Low	Poor	Small	High
Low	Poor	Medium	Medium
Low	Poor	Large	Low
Low	Average	NOT large	Medium
Low	Average	Large	Low
Low	Good	–	Low
Medium	Poor	NOT large	High
Medium	Poor	Large	Medium
Medium	Average	Small	Large
Medium	Average	Medium	Medium
Medium	Average	Large	Small
Medium	Good	Small	Medium
Medium	Good	NOT small	Low
High	Poor	–	High
High	Average	NOT large	High
High	Average	Large	Medium
High	Good	Small	High
High	Good	Medium	Medium
High	Good	Large	Small

TABLE 10.4

Fuzzy Inference Rules of Admission Processor

Required Resource	Cell Load	Decision
–	Light	Accept
NOT high	Moderate	Accept
High	Moderate	Weak accept
Low	Congested	Accept
Medium	Congested	Weak accept
High	Congested	Reject

the same subframe can be different. The PDU arrival process for a connection follows a two-state MMPP and the maximum batch size of arrival is 100 (i.e., $A = 100$).

The PDUs from a single traffic source are buffered into a queue at the mobile terminal. We assume that the queue size is 200 PDUs (i.e., $X = 200$) and it is same for all mobiles. For evaluation of queueing performance for a particular connection, we assume that the average SNR is 15 dB, and the MMPP parameters are given as follows:

$$\mathbf{M} = \begin{bmatrix} 0.8 & 0.2 \\ 0.8 & 0.2 \end{bmatrix} \quad \Lambda = \kappa \begin{bmatrix} 1 \\ & 5 \end{bmatrix} \tag{10.18}$$

where κ represents the traffic load corresponding to a traffic source and the arrival rate is given per frame. With these parameters, the probability of peak rate for a traffic source is 0.2 (i.e., $P_{peak} = 0.2$). We vary some of these parameters according to the evaluation scenarios, while the rest remain fixed.

For admission control, we assume that the minimum and the maximum required number of allocated subchannels for each connection is 1 and 8 (i.e., $c_{min} = 1$ and $c_{max} = 8$), respectively. We compare the proposed fuzzy admission control method with two other resource allocation and admission control schemes, namely, the static and the adaptive schemes. For the static scheme, the number of allocated subchannels is identical for every connection (e.g., $c_i = 5 \, \forall \, i$) and independent of traffic source parameters, delay requirement, and traffic load in a cell. In this scheme, an incoming connection is accepted if sufficient number of subchannels are available in the cell, and blocked otherwise. For the adaptive scheme, the number of allocated subchannels depends on the number of ongoing connections. Specifically, the admission control method tries to allocate the subchannels among the ongoing and the incoming connections equally. If the number of subchannels, which can be allocated to a connection is larger than or equal to the minimum number of subchannels c_{min}, then the incoming connection is accepted, and blocked otherwise. We also assume that an incoming connection is always blocked if the measured average SNR is below 7 dB to avoid very low transmission rate because of bad channel quality.

To evaluate the performance of the fuzzy admission control scheme, we consider a cell with radius of 5 km and the BS transmission power is 10 W for each subchannel. We assume that mobiles are uniformly distributed in the cell. For large-scale fading, we assume log-normal shadowing with standard deviation of 8 dB, path-loss exponent $n_l = 4$, reference distance $d_0 = 1$ km, and path-loss at reference distance $P_l(d_0) = 100$ dB. For small-scale fading, we assume a Nakagami-m fading channel with $m = 1.1$. With this setting, the average received SNR is in the range of 7–30 dB most of the time. Note that, this parameter setting is similar to that in Ref. 36.

10.7.2 Numerical and Simulation Results

10.7.2.1 Queueing Performances and Observations

Figures 10.10a and 10.10b show the average delay and the throughput performances, respectively, for a particular connection under different average SNR and number of allocated subchannels. As expected, the average delay decreases as the number of allocated subchannels increases. As the average SNR increases, the throughput increases and it approaches to a certain level, which corresponds to the maximum traffic arrival rate. Increasing the number of allocated subchannels results in faster increase in throughput, which implies that transmission rate is high enough to accommodate all arriving traffic. Similarly, better channel quality results in smaller delay and higher throughput. From these results, we observe that at poor average SNR (e.g., 7–10 dB), average delay is always large. When the channel quality becomes better (e.g., 10–18 dB), average delay decreases significantly. However, although the average SNR is high (e.g., 18–25 dB), the average delay slightly decreases. The three separate membership functions of average SNR as shown in Figure 10.8 are obtained based on this observation.

The average delay under different number of allocated subchannels and traffic load is shown in Figure 10.11. We observe that when the number of allocated subchannels is 1 or 2, the average delay is very high (i.e., >20 ms). However, when it is 3 or 4, the delay decreases substantially. Although allocating more number of subchannels results in smaller delay, the number of available subchannels for the other connections reduces as well. Consequently, only a fewer number of connections can be accommodated in the cell. The three membership functions for the number of allocated subchannels shown in Figure 10.8 are obtained based on this observation.

The average delay corresponding to different traffic source parameters (i.e., κ) and probability of peak rate are shown in Figures 10.12a and 10.12b, respectively. The average delay is observed to be less than 30 frames most of the time, except when the number of subchannels is small or the channel quality is poor (e.g., $c_i = 2$ and $\bar{\gamma} = 10\,dB$). As expected, the average delay increases as the probability of peak arrival rate increases. However, we observe that the average delay increases almost linearly with the probability of peak rate.

The rules for the resource-allocation processor are established based on the queueing performances. For example, fewer number of subchannels will be allocated to a connection with average channel quality when its delay requirement is large or the traffic intensity is low (Figure 10.12a). However, if the delay requirement is small, fewer number of subchannels will be allocated only if average SNR is high (Figure 10.10a).

10.7.2.2 Performances of Fuzzy Logic Admission Control

We assume that the connection arrival process in the cell follows Poisson distribution, and the mean arrival rate varies from 0.05 to 0.8 connections per minute and the connection holding time is exponentially distributed with an

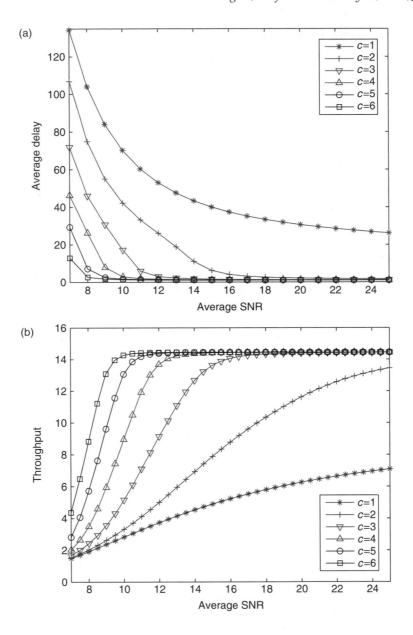

FIGURE 10.10
(a) Average delay and (b) throughput performances for a particular connection under different average SNR and a number of allocated subchannels.

average of 20 min. The normal and the peak rate for an incoming connection are assumed to be uniformly distributed in the range of 1–25 and 26–50 PDUs per frame, respectively. The probability of peak rate is between 0.1 and 0.5. The connection-level performances, that is, connection-blocking probability and average number of ongoing connections, are shown in Figures 10.13a and

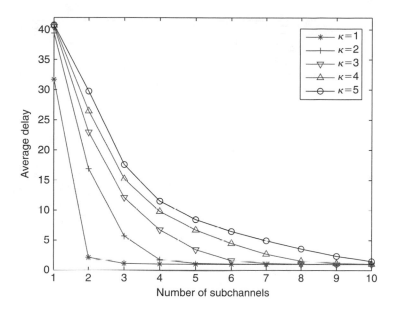

FIGURE 10.11
Average delay under different number of subchannels and traffic load.

10.13b, respectively. The average delay and the throughput performances for the ongoing connections are shown in Figures 10.14a and 10.14b, respectively.

We evaluate the performance of the proposed admission control in two scenarios with delay requirements of 5 and 10 frames ($d_{req} = 5$ and $d_{req} = 10$), which are indicated in both the figures by legends "fuzzy 1" and "fuzzy 2," respectively. Similarly, we evaluate the static admission control with the allocated number of subchannels being 4 and 5 (i.e., $c_i = 4$ and $c_i = 5$), which are indicated by legends "static 1" and "static 2," respectively.

As expected, the connection-blocking probability (Figure 10.13a) and the average number of ongoing connections (Figure 10.13b) increase as the connection arrival rate increases. However, the connection-blocking probability for each of the static and the fuzzy schemes is higher than that for the adaptive scheme. This is because of the fact that, in the static scheme, the number of connections is limited to 13 and 10 for the cases of $c_i = 4$ and $c_i = 5$, respectively, and the proposed fuzzy logic admission control limits the number of ongoing connections so that the delay requirement is satisfied. We observe that the average number of ongoing connections with the fuzzy admission control is slightly higher than that for the static scheme, especially in the case of "fuzzy 2" compared with "static 2," and therefore, the connection-blocking probability is smaller in the former case. However, the fuzzy scheme provides smaller delay (Figure 10.14a) and higher throughput (Figure 10.14b).

We also observe that with the adaptive admission scheme, even though the blocking probability is minimized (Figure 10.13a), it is unable to maintain

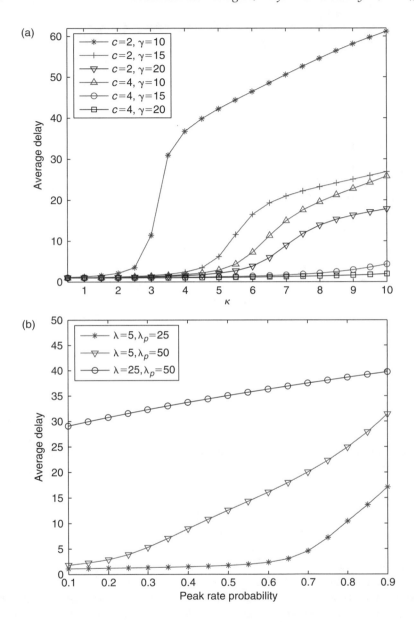

FIGURE 10.12
(a) Average delay under different traffic intensity and (b) different probabilities of peak rate.

the average PDU delay at the desired level. Specifically, when the traffic load in the cell increases, the performances (i.e., delay and throughput as shown in Figures 10.14a and 10.14b, respectively) degrade significantly. Therefore, we conclude that the proposed fuzzy logic admission control can increase the utilization of the available subchannels by considering the traffic

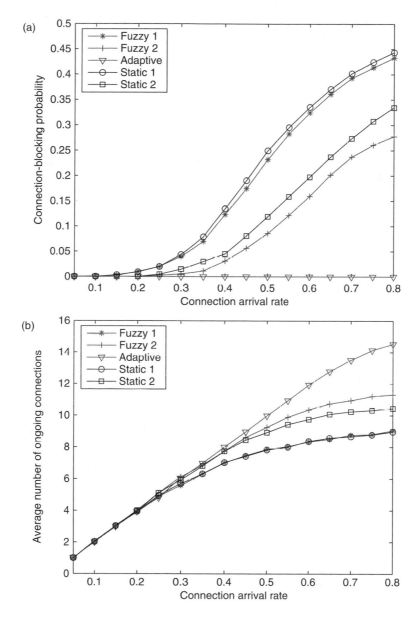

FIGURE 10.13
(a) Connection-blocking probability and (b) average number of ongoing connections.

arrival process while maintaining the delay requirement at the target level. Moreover, owing to the use of the admission inference processor, the QoS performances for the fuzzy scheme become insensitive to traffic load in the cell. Therefore, a mobile user has guaranteed QoS after the connection has been admitted.

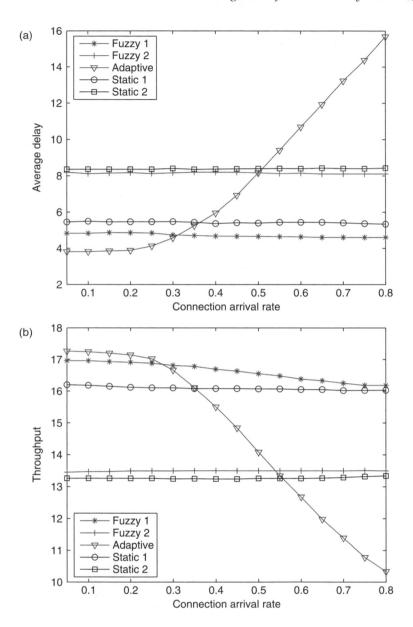

FIGURE 10.14
(a) Average delay and (b) throughput of ongoing connections.

The computation time required to obtain the number of allocated sub-channels and the admission decision based on the proposed scheme (i.e., for fuzzifying inputs, processing inference rules, and defuzzifying output) is observed to be less than 100 ms (using Matlab in a Pentium III 2.0 GHz PC with 512 MB memory).

10.8   Summary

We have proposed a fuzzy logic admission controller for OFDMA-based broadband wireless networks. The system under our consideration is compatible with the IEEE 802.16 in the TDD–OFDMA mode. The proposed algorithm is composed of three major components: traffic source estimator, resource allocation, and admission processor. These components take various system parameters such as the peak rate of traffic source, the channel quality, and the traffic load in the cell into account to estimate the traffic arrival intensity, allocate the available radio resources (i.e., subchannels), and admit/block new connection. We have also formulated a queueing model and utilized the packet-level QoS performance (i.e., average delay) obtained from the model to establish the inference rules for the resource-allocation processor. The performance of the proposed fuzzy admission control scheme has been evaluated by extensive simulations both in terms of packet-level (i.e., delay and throughput) and connection-level QoS measures (i.e., average number of connection and connection-blocking probability). We have also compared the performance of the proposed scheme with that for a static scheme and an adaptive scheme.

Acknowledgments

The authors acknowledge the support from TR*Labs* and Natural Sciences and Engineering Research Council (NSERC) of Canada.

References

1. IEEE 802.16 Standard—Local and Metropolitan Area Networks—Part 16, *IEEE Std 802.16a-2003*.
2. J. Ye, X. Shen, and J. W. Mark, Call admission control in wideband CDMA cellular networks by using fuzzy logic, *IEEE Trans. Mobile Comp.*, vol. 4, no. 2, pp. 129–141, March–April 2005.
3. S. Shen, C. J. Chang, C. Huang, and Q. Bi, Intelligent call admission control for wideband CDMA cellular systems, *IEEE Trans. Wireless Comm.*, vol. 3, no. 5, pp. 1810–1821, September 2004.
4. D. Niyato and E. Hossain, Delay-based admission control using fuzzy logic for OFDMA broadband wireless networks, *Proc. IEEE ICC'06*, vol. 12, pp. 5511–5516, June 2006.
5. Y.-C. Liang, R. Zhang, and J. M. Cioffi, Transmit optimization for MIMO-OFDM with delay-constrained and no-delay-constrained traffic, *IEEE Trans. Signal Process.*, vol. 54, no. 8, pp. 3190–3199, August 2006.

6. I. Koutsopoulos and L. Tassiulas, Cross-layer adaptive techniques for throughput enhancement in wireless OFDM-based networks, *IEEE/ACM Trans. Network.*, vol. 14, pp. 1056–1066, October 2006.

7. I. Kim, I.-S. Park, and Y. H. Lee, Use of linear programming for dynamic subcarrier and bit allocation in multiuser OFDM, *IEEE Trans. Vehicular Technol.*, vol. 55, no. 4, pp. 1195–1207, July 2006.

8. J. Bandara, X. Shen, and Z. Nurmohamed, A fuzzy resource controller for non-real-time traffic in wireless networks, *Proc. IEEE ICC'00*, vol. 1, pp. 75–79, June 2000.

9. J. Mar and C.-Y. Kao, Cascade fuzzy radio resource management using SDMA scheduling in TD-CDMA system, *Proc. IEEE WCNC'06*, vol. 2, pp. 991–996, April 2006.

10. L. Giupponi, R. Agusti, J. Perez-Romero, and O. Sallent, A novel joint radio resource management approach with reinforcement learning mechanisms, *Proc. IEEE IPCCC'05*, pp. 621–626, April 2005.

11. L. Giupponi, R. Agusti, J. Perez-Romero, and O. Sallent, Joint radio resource management algorithm for multi-RAT networks, *Proc. IEEE GLOBECOM'05*, vol. 6, November–December 2005.

12. D. Soud and H. B. Kazemian, A fuzzy approach to connection admission control in 802.11b wireless systems, *Proc. IEEE IECON'03*, vol. 2, pp. 1914–1919, November 2003.

13. C.-L. Chen and P.-C. Hsiao, Supporting QoS in wireless MAC by fuzzy control, *Proc. WCNC'05*, vol. 2, pp. 1242–1247, March 2005.

14. A. Zhu and J. Hu, Adaptive call admission control for multi-class CDMA cellular systems, *Proc. IEEE APCC/OECC'99*, vol. 1, pp. 533–536, October 1999.

15. B. AliPanahi and M. Karzand, A novel soft handoff algorithm for fair network resources distribution, *Proc. IEEE WOCN'05*, pp. 255–259, March 2005.

16. G. Edwards and R. Sankar, Hand-off using fuzzy logic, *Proc. IEEE GLOBECOM'95*, vol. 1, pp. 524–528, November 1995.

17. G. Edwards and R. Sankar, A predictive fuzzy algorithm for high performance microcellular handoff, *Proc. IEEE GLOBECOM'97*, vol. 2, pp. 987–990, November 1997.

18. M. S. Dang, A. Prakash, D. K. Anvekar, D. Kapoor, and R. Shorey, Fuzzy logic based handoff in wireless networks, *Proc. IEEE VTC'00 Spring*, vol. 3, pp. 2375–2379, May 2000.

19. N. D. Tripathi, J. H. Reed, and H. F. VanLandingham, Fuzzy logic based adaptive handoff algorithms for microcellular systems, *Proc. IEEE VTC'99*, vol. 2, pp. 1419–1427, May 1999.

20. H. Liao, L. Tie, and Z. Du, A vertical handover decision algorithm based on fuzzy control theory, *Proc. IEEE IMSCCS'06*, vol. 2, pp. 309–313, April 2006.

21. J. Hou and D. C. O'Brien, Vertical handover-decision-making algorithm using fuzzy logic for the integrated Radio-and-OW system, *IEEE Trans. Wireless Commn.*, vol. 5, no. 1, pp. 176–185, January 2006.

22. P. M. L. Chan, R. E. Sheriff, Y. F. Hu, P. Conforto, and C. Tocci, Mobility management incorporating fuzzy logic for heterogeneous a IP environment, *IEEE Commn. Mag.*, vol. 39, no. 12, pp. 42–51, December 2001.

23. W. Zhang, Handover decision using fuzzy MADM in heterogeneous networks, *Proc. IEEE WCNC'04*, vol. 2, pp. 653–658, March 2004.

24. Q. Guo, J. Zhu, and X. Xu, An adaptive multi-criteria vertical handoff decision algorithm for radio heterogeneous network, *Proc. IEEE ICC'05*, vol. 4, pp. 2769–2773, May 2005.

25. Y. Nkansah-Gyekye and J. I. Agbinya, Vertical handoff between WWAN and WLAN, *Proc. IEEE ICN/ICONS/MCL'06*, p. 132, April 2006.

26. K. Murray and D. Pesch, Intelligent network access and inter-system handover control in heterogeneous wireless networks for smart space environments, *Proc. IEEE ISWCS'04*, pp. 66–70, September 2004.

27. T. Kwon, H. Lee, S. Choi, J. Kim, D.-H. Cho, S. Cho, S. Yun, W.-H. Park, and K. Kim, Design and implementation of a simulator based on a cross-layer protocol between MAC and PHY layers in a WiBro Compatible IEEE 802.16e OFDMA system, *IEEE Commn. Mag.*, vol. 43, no. 12, pp. 136–146, December 2005.

28. M. Zorzi, PDU dropping statistics of a data-link protocol for wireless local communications, *IEEE Trans. Vehicular Technol.*, vol. 52, no. 1, pp. 71–79, January 2003.

29. Q. Liu, S. Zhou, and G. B. Giannakis, Queueing with adaptive modulation and coding over wireless links: cross-layer analysis and design, *IEEE Trans. Wireless Commn.*, vol. 4, no. 3, pp. 1142–1153, May 2005.

30. D. Niyato and E. Hossain, A queueing-theoretic and optimization-based model for radio resource management in IEEE 802.16 broadband wireless networks, *IEEE Trans. Comput.*, vol. 55, no. 11, pp. 1473–1488, November 2006.

31. D. Niyato and E. Hossain, Queue-aware uplink bandwidth allocation and rate control for polling service in IEEE 802.16 broadband wireless networks, *IEEE Trans. Mobile Computing*, vol. 5, no. 6, pp. 668–679, June 2006.

32. L. Muscariello, M. Meillia, M. Meo, M. A. Marsan, and R. L. Cigno, An MMPP-based hierarchical model of Internet traffic, *Proc. IEEE ICC'04*, vol. 4, pp. 2143–2147, June 2004.

33. P. Salvador, R. Valadas, and A. Pacheco. Multiscale fitting procedure using Markov modulated Poisson processes, *Telecommn. Syst.*, vol. 23, pp. 123–148, 2003.

34. C. D. Meyer, *Matrix Analysis and Applied Linear Algebra*, SIAM, Philadelphia, PA, 2000.

35. D. Dubois and H. Prade, *Fuzzy Sets and Systems: Theory and Applications*, Academic Press, New York, 1980.

36. I. Koffman and V. Roman, Broadband wireless access solutions based on OFDM access in IEEE 802.16, *IEEE Commn. Mag.*, vol. 40, no. 4, pp. 96–103, April 2002.

Index

A

adaptive antenna systems (AASs), space-time coding applications, 43–44

admission control
    fuzzy logic, 250–254, 257
    IEEE 802.16 standard, 98–99

Alamouti Code
    basic properties, 50–52
    differential coding, 55–56
    orthogonal frequency division multiplexing
        numerical and simulation results, 163–168
        space-time block codes, 156–163

analog-to-digital (ADC) converter, WiMax system components, 4–5

antenna systems, MIMO-OFDM systems, system model, 72–73

application-specific instruction set processors (ASIPS), 8–10

approximation plots, fitting algorithms, 105–107

automatic gain control (AGC) circuits, baseband processor noise and burst interference, 7

automatic repeat request (ARQ)-based error control, queueing formalism, 245

B

bandwidth properties, traffic throughput classes, 107–108

bandwidth request (BWR), reservation-based multiple access control (R-MAC) protocol, 179–181
    Markov decision process (MDP) model, 189–190
    performance evaluation, 182–183
    p-persistence contention resolution, 201–207

baseband processors
    dynamic MIPS allocation, 8–9
    dynamic range problems, 6–7
    hardware multiplexing, 9–10
    IEEE 802.16d example, 10–13
    latency, 7

mobility, 56

multimode systems, 8

multipath propagation, 5–6

multistandard processor design, 13–19
    complex computing, 13
    execution units, 15–16
    FFT acceleration, 17–18
    forward error correction, 18–19
    front-end acceleration, 18
    hardware acceleration, 17
    LeoCore processor, 14–17
    memory subsystem, 16–17
    single instruction issues, 15
    vector computing, 13–14

noise and burst interference, 6–7

programmable system, 7–8

timing and frequency offset, 5

WiMax system components, 4–5

Bell Labs Layered Space-Time Architecture (BLAST) framework, space-time block codes, spatial multiplexing, 50

best effort (BE) services
    FIFO/TXOP allocation, 109, 111
    fitting algorithms, 106–107
    IEEE 802.16 standard, 98
    nrTPS traffic load vs., 111, 113
    orthogonal frequency division multiplexing
        numerical and simulation results, 163–168
        single-input-single-output performance, 149–156
        space-time block coding, 156–163
    performance evaluation, 108–109
    quality-of-service (QoS) features, 100
    reservation-based multiple access control (R-MAC) protocol, 175–176, 179
    throughput vs. traffic load, 113, 115

bit-error rate (BER)
    MIMO-OFDM systems, 86–91
    orthogonal frequency division multiplexing, single-input-single- output systems, 154–156
    WiMAX forum, 70